Urban Observation and Social Sensing

Urban Observation and Social Sensing helps readers see cities in a way they have seldom imagined before. Advanced technologies such as satellites in space, near surface drones, or even smartphones, can be used to explain how a city functions, or why certain areas are more populated than others. It examines the science of capturing city images and information gathered from above the ground combined with the information from social media and mobile data on the ground, and how to make sense of it all. From planning better neighborhoods to understanding how city life affects health, this book is a guide to the future of urban living and the technologies that help navigate it.

Features

- Covers a broad spectrum of urban observation methods in one resource to ensure a thorough understanding of diverse data collection methods.
- Provides a unique emphasis on social sensing to enable a more holistic and human-centric view of urban landscapes.
- Includes practical applications to showcase how the data and methodologies can be used in real-world scenarios.
- Explains the current technologies and methodologies but also offers insights into the future trajectory of urban science.
- Integrates technological advances with informed urban planning and real-time insights to elucidate a holistic understanding of urban spaces.

This is an excellent, insightful resource for researchers, academics, students, and professionals in remote sensing, Geographic Information Systems, urban planning and design, data science, geography, earth science, and those interested in the future of urban living and planning.

The Urban-Environment-Sustainability-Technology Nexus
Series Editor: Qihao Weng, Chair Professor, Department of Land Surveying and Geo-Informatics, Hong Kong Polytechnic University, Hung Hom, Kowloon, Hong Kong

This book series examines global urbanization's environmental impacts across different geographical settings and development stages. Rapid worldwide urbanization has dramatically altered ecosystems, biodiversity, climate, and water resources, creating complex challenges for assessment and monitoring. While urbanization generates local effects, evidence shows significant implications for regional environments and global climate change. The series addresses sustainable urban development strategies across continents, involving interdisciplinary research from environmental science, urban studies, data science, engineering, and social theories. Key topics include green cities, resilient cities, smart cities, urban computing, land use planning, climate, ecosystems, and geospatial AI.

Urban Observation and Social Sensing
Edited by Bin Chen and Fan Zhang

Urban Observation and Social Sensing

Edited by

Bin Chen and Fan Zhang

CRC Press
Taylor & Francis Group
Boca Raton London New York

CRC Press is an imprint of the
Taylor & Francis Group, an **informa** business

Designed cover image: © Shutterstock

First edition published 2027
by CRC Press
2385 NW Executive Center Drive, Suite 320, Boca Raton FL 33431

and by CRC Press
4 Park Square, Milton Park, Abingdon, Oxon, OX14 4RN

CRC Press is an imprint of Taylor & Francis Group, LLC

Library of Congress Cataloging-in-Publication Data
Names: Chen, Bin (Associate professor) editor | Zhang, Fan (Assistant professor) editor
Title: Urban observation and social sensing / edited by Bin Chen and Fan Zhang.
Description: First edition. | Boca Raton FL : CRC Press, 2026. | Series: Urban-environment-sustainability-technology nexus | Includes bibliographical references.
Identifiers: LCCN 2025053504 (print) | LCCN 2025053505 (ebook) | ISBN 9781032776538 hardback | ISBN 9781032776569 paperback | ISBN 9781003484219 ebook
Subjects: LCSH: Cities and towns--Remote sensing | City planning--Remote sensing | Environmental monitoring--Remote sensing
Classification: LCC HT165.5 .U726 2026 (print) | LCC HT165.5 (ebook)
LC record available at https://lccn.loc.gov/2025053504
LC ebook record available at https://lccn.loc.gov/2025053505

ISBN: 978-1-032-77653-8 (hbk)
ISBN: 978-1-032-77656-9 (pbk)
ISBN: 978-1-003-48421-9 (ebk)

DOI: 10.1201/9781003484219

Typeset in Palatino
by KnowledgeWorks Global Ltd.

For Product Safety Concerns and Information please contact our
EU representative GPSR@taylorandfrancis.com Taylor & Francis
Verlag GmbH, Kaufingerstraße 24, 80331 München, Germany

Contents

Preface

The idea for this book grew out of a growing discomfort shared by many in our field: while cities are transforming faster than at any point in human history, our tools for observing and understanding them have not kept pace with the complexity of the challenges they face. Urban heat crises, pandemics, aging infrastructure, biodiversity loss, energy stress, informal settlement growth, and widening socio-spatial inequities are no longer isolated phenomena—they are interlinked, accelerating, and deeply spatial. Yet much of urban research still relies on fragmented data, coarse generalizations, or disciplinary silos. Over years of working with satellite observations, AI, public health data, urban morphology, and social sensing, we realized that the most important stories about cities were falling through disciplinary gaps. We needed a new vocabulary and new methodological frameworks that integrate physical environments, human behaviors, and policy-relevant insights. That realization—and the urgency of climate change, demographic shifts, and digital disruptions—inspired this volume.

This book brings together the diverse strands of urban observations not as parallel developments, but as parts of a single evolving paradigm. The chapters reflect how the field has moved far beyond land cover detection and mapping. We now extract urban form from multispectral data, trace socioeconomic activity from nighttime lights, model material properties using microwave signals, reconstruct forests and building volumes with LiDAR, and monitor urban change in real time through nanosatellite constellations. At the same time, the boundary of "urban observation" itself has expanded. Street-view imagery, social media platforms, volunteered geographic information, and navigation patterns offer windows into how environments are perceived, used, and experienced. These emerging data sources force us to rethink what counts as urban observation and who participates in it.

What makes this book timely—and necessary—is that it does not treat these technologies as isolated tools but as interconnected instruments for reimagining urban research. From human thermal exposure in informal housing to AI-enabled facility planning, from spatial cognition and navigation to the geometry of living structures, each chapter addresses a dimension of urban life that traditional sensing could not reach. Taken together, they show how environmental sensing, behavioral insights, and spatial intelligence can converge to confront some of the most pressing urban challenges of this century. This volume is for researchers, educators, practitioners, and students who believe that cities must be understood as dynamic, data-rich, and human-centered systems. It is also a call for the next generation of scholars to move beyond observational boundaries and disciplinary comfort zones. The future of urban science will not be built on better sensors alone, but on better questions, deeper integration, and a willingness to see cities as living, sensing, learning entities. We hope this book contributes to that shift—not only by synthesizing progress, but by challenging assumptions about what urban observation and sensing can be and whom it should serve.

Bin Chen and Fan Zhang
20 October 2025

About the Editors

Dr. Bin Chen is an Associate Professor, Director of Future Urbanity & Sustainable Environment (FUSE) Lab at The University of Hong Kong. He obtained his BSc in Geographical Information Systems from Wuhan University, and his PhD in Global Environmental Change from Beijing Normal University. Before joining the University of Hong Kong in 2021, he worked as a postdoctoral researcher at the University of California, Davis.

Dr. Chen's research leverages geospatial big data, data-model fusion, and advanced interdisciplinary approaches to investigate the interaction loops between environmental change, human activities, and public health. Specifically, his major research includes the remote sensing of built and natural environmental changes; modelling of human-environment spatiotemporal interaction; and impact of environment and human activities on public health. He has published over 80 refereed publications in top academic journals. His achievements have received international recognition, including Geospatial World 50 Rising Stars (2024) and receiving the International Society of Digital Earth Young Scientist Award, American Association of Geographers Early Career Award in Remote Sensing, and HKU-100 Scholars Award and Outstanding Young Researcher Award.

Dr. Fan Zhang is an Assistant Professor at the Institute of Remote Sensing and GIS, Peking University. Prior to this, he was a Senior Research Fellow at MIT, where he led the Urban Visual AI group at the MIT Senseable City Lab. His research focuses on geographical artificial intelligence and data-driven urban studies, with particular emphasis on developing urban visual intelligence using street-level imagery. Dr. Zhang's work has been published in renowned academic journals, and he currently serves as an Associate Editor and Editorial Board Member for various journals. He currently serves as an Associate Editor, an Editorial Board Member. His notable honors include the CPGIS Young Scholar Award (2024), the Geospatial World 50 Rising Stars Award (2022), and the Global Young Scientist Award in Frontier Science and Technology from WGDC (2020).

Contributors

Ronita Bardhan
Sustainable Design Group
University of Cambridge
Cambridge, UK

Bin Cheng
Future Urbanity and Sustainable Environment (FUSE) Lab
Division of Landscape Architecture, Department of Architecture
Faculty of Architecture
The University of Hong Kong
Pok Fu Lam, Hong Kong

Bisong Hu
School of Geography and Environment
Jiangxi Normal University
Nanchang, China

Furong Jia
The Institute of Remote Sensing and Geographical Information Systems
School of Earth and Space Sciences
Peking University
Beijing, China

Bin Jiang
LivableCityLAB
Urban Governance and Design Thrust
Hong Kong University of Science and Technology
Guangzhou, China

Ziming Li
Future Urbanity and Sustainable Environment (FUSE) Lab
Division of Landscape Architecture, Department of Architecture
Faculty of Architecture
The University of Hong Kong
Pok Fu Lam, Hong Kong

Rui Liu
Department of Geography
The University of Hong Kong
Pokfulam, Hong Kong

Jiayu Pan
Sustainable Design Group
University of Cambridge
Cambridge, UK

Huan Qian
LivableCityLAB
Urban Governance and Design Thrust
Hong Kong University of Science and Technology
Guangzhou, China

Xinyi Shen
School of Resource and Environmental Sciences
Wuhan University
Wuhan, China

Liqun Sun
Shenzhen Institute of Advanced Technology
Chinese Academy of Sciences
Shenzhen, China

Maoran Sun
Sustainable Design Group
University of Cambridge
Cambridge, UK

Han Wang
Future Urbanity and Sustainable Environment (FUSE) Lab
Division of Landscape Architecture, Department of Architecture
Faculty of Architecture
The University of Hong Kong
Pok Fu Lam, Hong Kong

Shaohua Wang
Key Laboratory of Remote Sensing and Digital Earth, Chinese Academy of Sciences
Aerospace Information Research Institute
Chinese Academy of Sciences
Beijing, China
And
International Research Center of Big Data for Sustainable Development Goals, CAS
Beijing, China

Xiaoling Wang
Future Urbanity and Sustainable Environment (FUSE) Lab
Division of Landscape Architecture, Department of Architecture
Faculty of Architecture
The University of Hong Kong
Pok Fu Lam, Hong Kong

Yuqing Wang
The Institute of Remote Sensing and Geographical Information Systems
School of Earth and Space Sciences
Peking University
Beijing, China

Gang Xu
School of Resource and Environmental Sciences
Wuhan University
Wuhan, China

Qianxiang Yao
LivableCityLAB, Urban Governance and Design Thrust
Hong Kong University of Science and Technology
Guangzhou, China

Fan Zhang
The Institute of Remote Sensing and Geographical Information Systems
School of Earth and Space Sciences
Peking University
Beijing, China

Hongsheng Zhang
Department of Geography
The University of Hong Kong
Pokfulam, Hong Kong

Longrui Zhu
School of Resource and Environmental Sciences
Wuhan University
Wuhan, China

1

Multispectral Remote Sensing in Urban Environments

Ziming Li and Han Wang

1.1 Overview

Remote sensing refers to remotely capture information via multiple sensors without any physical contact. This technology is widely used in the field of Earth observation. Typically using satellites or aircrafts, remote sensing can gather information about the Earth's surface by detecting and measuring emitted or reflected electromagnetic radiation from objects and areas on Earth. Remote sensing has numerous important applications, including environment, earth science, geography, and social science.

The fundamental principle of remote sensing is the interaction between electromagnetic radiation and the Earth's surface or atmosphere. Different objects and materials absorb, reflect, or emit electromagnetic radiation uniquely, producing distinct spectral signatures. The spectral signal intensity varies depending on the object's physical and chemical properties, allowing for differentiation and classification of diverse surface features.

From the perspective of spectral resolution, remote sensing can be classified into three categories: panchromatic, multispectral, and hyperspectral. Spectral resolution is a key factor in distinguishing between panchromatic, multispectral, and hyperspectral remote sensing. Panchromatic sensors have low spectral resolution as they capture data across a broad wavelength range, typically encompassing the entire visible spectrum (approximately 400–700 nm) or even extending into the near-infrared region. This results in a single-band grayscale image where each pixel represents the intensity of reflected light across the broad wavelength range, making it useful for detailed mapping and analysis of urban areas and infrastructure due to its high spatial resolution. On the other hand, multispectral sensors offer moderate spectral resolution by capturing data in several discrete spectral bands, typically ranging from 3 to 15 bands, covering specific portions of the electromagnetic spectrum including visible, near-infrared (NIR), and sometimes shortwave infrared (SWIR) regions (approximately 400–2500 nm). This results in a set of images, each corresponding to a specific spectral band,

DOI: 10.1201/9781003484219-1

which can be combined to form color composites like true color and false color images for various purposes. Multispectral imaging is widely used for vegetation analysis, land cover classification, water quality assessment, and environmental monitoring due to its ability to effectively discriminate broad categories of surface materials and conditions. In contrast, hyperspectral sensors provide high spectral resolution by capturing data in a large number of narrow and contiguous spectral bands, often exceeding 100 bands, spanning a wide range of the electromagnetic spectrum from the visible to NIR and SWIR. Some advanced systems can further extend into the mid-infrared (MIR) region (above 3000 nm).This results in a detailed spectral signature for each pixel, allowing for precise material identification and quantification. Hyperspectral imaging is used in applications that require detailed material identification and analysis, such as mineral exploration, precision agriculture, environmental monitoring, and medical diagnostics. The high spectral resolution enables the detection of subtle differences in spectral signatures, making it possible to identify specific materials and conditions that multispectral and panchromatic sensors cannot resolve. In summary, panchromatic sensors offer high spatial but low spectral resolution, making them suitable for detailed spatial analysis but limited in material discrimination. Multispectral sensors provide a balance between spatial and spectral resolution, allowing for effective broad-category discrimination and environmental monitoring. Hyperspectral sensors deliver high spectral resolution, enabling detailed material identification and analysis, though often with lower spatial resolution compared to panchromatic sensors. Spectral resolution thus serves as the defining characteristic that differentiates these three types of remote sensing, each with its specific strengths and applications.

In this chapter, we primarily focus on the multispectral remote sensing technique. Multispectral remote sensing is characterized as using sensors to capture information within several narrow bands of the electromagnetic spectrum, primarily including visible and NIR bands. Most multispectral images contain 3 to 15 bands ranging from around 400 nm to 2500 nm. The specific wavelengths of these bands can vary depending on the instruments being used. Table 1.1 demonstrates common spectral bands and their typical applications.

To the simulate human vision system and enhance imaging contrast and detail, it is a common practice to select three specific bands to form RGB composites, thereby supporting multispectral remote sensing analysis. This approach leverages the natural sensitivity of the human eye to red, green, and blue light, creating images that are both intuitive to interpret and rich in information. By combining these three bands, remote sensing imagery can effectively highlight various features and materials on the Earth's surface, making it easier to distinguish between different types of vegetation, water bodies, urban structures, and geological formations. Table 1.2 shows typical three-band combinations and their applications.

TABLE 1.1

Classical Spectral Bands Used in Multispectral Remote Sensing

Spectral Band	Wavelength Range (nm)	Applications
Blue	450–515/520	Atmosphere and deep water imaging
Green	515/520–600	Vegetation and deep water structures
Red	600/630–690	Man-made objects, water bodies, soil, and vegetation
Near infrared (NIR)	750–1400	Vegetation imaging
Shortwave infrared (SWIR)	1400–3000	Moisture content of soil and vegetation, minerals, and rock types
Mid infrared (MIR)	3000–8000	Vegetation, soil moisture content, and some forest fires
Thermal infrared (TIR)	8000–15000	Imaging soil, moisture, geological features, silicates, clays, and fires
Far infrared (FIR)	15000–1000000	Atmospheric and astronomical research

TABLE 1.2

Common Composite of Multispectral Bands

Composite Type	Bands (Wavelength Range)	Applications
True color	Red Green Blue	Mimics natural human vision and is used for general landscape and urban area mapping.
False color	NIR Red Green	Vegetation appears in shades of red, making it easier to distinguish from other features; useful for assessing plant health and vegetation cover.
Agricultural	NIR Red Blue	Effective for monitoring crop health and vigor, as healthy vegetation reflects more NIR and less visible light.
Geological	SWIR 1 NIR Red	Used for identifying minerals and geological formations. SWIR bands are particularly useful for distinguishing soil and rock types.
Urban	SWIR 1 NIR Blue	Useful for urban studies, as it highlights built-up areas and infrastructure; helps in differentiating between vegetation, water, and urban features.
Water body analysis	SWIR Red Green	Ideal for mapping water bodies and assessing water quality; NIR and SWIR bands are sensitive to water content and can help in detecting moisture levels.

These classical three-band combinations are exemplary uses of multispectral remote sensing data for visual interpretation. By selecting appropriate bands, it is possible to enhance the visibility of specific features and gain valuable insights into various environmental and man-made phenomena.

Globally, numerous organizations are involved in the development and maintenance of various remote sensing platforms. Table 1.3 shows a range of widely used remote sensing sensors and platforms, operated by organizations such as NASA, the USGS, the European Space Agency (ESA), Maxar Technologies, and Planet Labs. These platforms offer various spatial resolutions and spectral bands, catering to diverse applications.

Multispectral remote sensing offers several advantages that make it a valuable tool for a wide range of applications. It provides broad geographic coverage, capturing data over large areas essential for regional and global studies. The use of multiple discrete spectral bands allows for the detailed

TABLE 1.3

Multispectral Remote Sensing Sensors and Platforms

Platform	Operator	Bands	Spatial Resolution	Applications
Landsat	NASA and USGS	11 bands (visible, near-infrared, shortwave infrared)	30 m	Land use and land cover mapping, agriculture, forestry, environmental monitoring, urban planning
Sentinel-2	European Space Agency (ESA)	13 bands (visible, near-infrared, shortwave infrared)	10 to 60 m	Agriculture, forestry, land use, water quality monitoring, disaster management
MODIS	NASA	36 bands	250 m 500 m 1 km	Large-scale environmental and climate monitoring, vegetation health, land surface temperature, ocean color
WorldView-3	Maxar Technologies	16 bands (visible, near-infrared, shortwave infrared)	Up to 1.24 m	Precision agriculture, urban planning, infrastructure monitoring, environmental assessment
PlanetScope	Planet Labs	4–8 bands (red, green, blue, NIR)	3–5 m	Agriculture, forestry, land use monitoring, disaster response

analysis of different surface materials and conditions, enabling the distinction between various types of vegetation, water bodies, soil, and urban features. Regular, repeatable observations from satellite-based multispectral sensors facilitate effective temporal monitoring, crucial for monitoring environmental changes such as seasonal vegetation growth, deforestation, urban expansion, and the impact of natural disasters. Additionally, multispectral remote sensing is generally more cost-effective than hyperspectral imaging, making it accessible for diverse applications and users. The data can be easily integrated with other geospatial information, enhancing comprehensive analysis and informed decision-making. Its versatility spans multiple fields, including agriculture, forestry, urban planning, environmental monitoring, and disaster management. The ability to create true color and false color composites enhances visual interpretation, aiding in the identification and analysis of specific features and conditions. Furthermore, capturing data in multiple spectral bands improves the accuracy of classification and analysis, allowing for precise differentiation of vegetation types and assessment of water quality parameters.

1.2 Case Studies

Using remote sensing technology to observe, monitor, and analyze urban environments is highly impactful. Remote sensing in an urban context involves leveraging spectral, texture, structural, and spatial features to identify and analyze urban environments; integrating image data with relational and statistical analysis to address socioeconomic phenomena; applying remote sensing techniques to urban-specific resource, ecological, and environmental challenges; providing essential information for urban planning, assessment, and predictive modeling; and enabling a comprehensive understanding of urban dynamics by combining remote sensing insights with other urban processes. Multispectral data with the advantages in the high accessibility, availability, and informativity, have facilitated a broad range of urban applications, which can be divided into two categories: urban object detection and environmental monitoring.

1.2.1 Urban Object Extraction and Classification

Urban environments are inherently complex and heterogeneous, exhibiting significant challenges for accurate object detection and classification in urban areas. Unlike the rural or natural landscapes, where homogeneous land covers can be more easily classified, urban areas require improved approaches and techniques to effectively excavate and interpret the spatial patterns and spectral characteristics of the built environment. Traditional ground-based survey methods, while accurate and detailed, are often impractical for large-scale urban analysis due to their labor-intensive nature and the rapidly changing dynamics of urban spaces. In contrast, multispectral remote sensing offers a comprehensive view and rich information, enabling the efficient monitoring of urban expansion, land use changes, and environmental impacts on a broader scale. Therefore, medium- to high-resolution remote sensing data has become indispensable for urban studies, particularly in the context of multispectral imaging. Medium- to high-resolution multispectral imagery is essential for capturing the fine-grained details needed to differentiate between complex urban features. In the following sections, we will delve into the methodologies used for extracting and classifying key urban objects: urban greenspaces (UGSs), water bodies, buildings, and roads, demonstrating how multispectral remote sensing data can be effectively used in these tasks.

1.2.1.1 Urban Greenspace

UGSs, including parks, gardens, urban forests, residential greenery, and green roofs, are important components of the urban ecosystem. These spaces offer a range of essential ecosystem services, such as air purification, temperature regulation, and recreational areas for residents, all of which contribute to improving environmental quality and human well-being. Consequently, the accurate identification and classification of UGSs are critical for effective urban planning, environmental protection, and biodiversity conservation within urban areas.

The distinct spectral characteristics of UGSs are the foundation of designing feasible methodologies for classification with multispectral remote sensing data. The reflectance spectrum of vegetation is shaped by the physical and biochemical properties of the plant leaves and canopies, with spectral bands revealing various aspects of vegetation. In the visible spectrum (400–760 nm), the reflectance rate is mainly influenced by the chlorophyll in plant leaves, leading to a small reflectance peak around 550 nm (green), accompanied by absorption bands near 450 nm (blue) and 670 nm (red), which means chlorophyll absorbs blue and red light strongly while reflecting green light. In the NIR region (700–1100 nm), the reflectance of vegetation exhibits a substantial increase, which is known as the "red edge". This feature is primarily due to the internal structure of plant leaf cells, which efficiently scatter NIR radiation, resulting in high reflectance. In the SWIR region (1300–2500 nm), reflectance is heavily influenced by the water content within. Here, the absorption increases, leading to lower reflectance, particularly at the water absorption bands centered around 1450 nm, 1950 nm, and 2700 nm. These spectral characteristics not only enable the differentiation of vegetated areas from non-vegetated surfaces in urban environments but also allow for the classification of different species and the assessment of vegetation health and environmental conditions.

such as discontinuous roads and irregular building boundaries. To address this, a post-processing step is introduced using Simple Linear Iterative Clustering (SLIC) to segment the CNN output into superpixels. These superpixels are then structured as Region Adjacency Graph (RAG), which identifies adjacent regions for potential merging or splitting. To distinguish roads and buildings, specific shape features are used: elongation and asymmetry for roads, and compactness and density for buildings. Roads are typically elongated and narrow, and regions with high elongation and asymmetry are more likely to be classified as roads. In contrast, buildings tend to be compact and dense, so regions with high compactness and density are classified as buildings. These low-level shape features are integrated with CNN features to connect disjointed road regions and complete irregular boundaries of building regions. The proposed method is evaluated on two datasets: the Massachusetts dataset and the Abu Dhabi dataset, showing significant improvements in maintaining road connectivity and ensuring precise building edge delineation.

1.2.2 Urban Environment Monitoring

Urban environmental issues are increasingly severe, significantly impacting the well-being of urban residents. These challenges include air and water pollution, urban heat islands (UHIs), and the loss of green spaces, all of which detrimentally affect public health and quality of life. Effective management of these issues necessitates advanced observation technologies for comprehensive monitoring and analysis. Multispectral remote sensing technology offers significant potential for addressing urban environmental challenges through its advanced observation and analysis capabilities. This technology, leveraging multiple spectral bands, offers a powerful approach for spatio-temporal analysis of urban environmental dynamics. The utilization of diverse spectral ranges enables the simultaneous capture of various environmental parameters, facilitating comprehensive urban ecosystem assessment. Specifically, the integration of visible, NIR, and TIR bands allows for detailed characterization of urban land cover, vegetation health, surface temperature distributions, and air quality indices. This multi-band approach enhances the ability to discriminate between different urban features and environmental conditions with high spatial resolution.

1.2.2.1 Atmospheric Pollution

Urban air pollution encompasses a variety of contaminants, including soot, total suspended particulates (TSP), inhalable particles (PM10), fine particulate matter ($PM_{2.5}$), nitrogen dioxide (NO_2), sulfur dioxide (SO_2), carbon monoxide (CO), ozone (O_3), and volatile organic compounds (VOC). These pollutants pose significant threats to public health, contributing to a range of diseases. Exposure to fine $PM_{2.5}$ and NO_2 is strongly associated with respiratory conditions such as asthma, bronchitis, and chronic obstructive pulmonary disease. Long-term exposure to high levels of $PM_{2.5}$ and O_3 can lead to cardiovascular diseases, including heart attacks, hypertension, and stroke. SO_2 and CO are linked to reduced lung function and exacerbation of existing heart conditions. Additionally, VOCs are known to be carcinogenic, increasing the risk of cancers such as lung cancer. The cumulative effects of these pollutants can also impair immune function, leading to increased susceptibility to infections.

Multispectral remote sensing utilizes various wavelength bands to effectively monitor specific air pollutants. UV bands are sensitive to O_3 absorption in the 280–320 nm range and SO_2 in the 300–400 nm range. Visible bands, ranging from 400–700 nm, are effective for detecting TSP and PM10 due to their scattering and absorption of visible light, and NO_2 with distinct absorption features in the 400–450 nm range. NIR bands, particularly around 700–1100 nm and 850–900 nm, are sensitive to $PM_{2.5}$ and some VOCs. MIR bands, in the 2.3–2.5 micrometers range, detect CO, utilized by instruments like the Atmospheric Infrared Sounder, and CO_2 around 4.3 micrometers. TIR bands are sensitive to SO_2 in the 8–14 micrometers range, particularly around 8.6 micrometers, and O_3 around 9.6 micrometers. Microwave bands, at 22.235 GHz and 183.31 GHz, are highly sensitive to water vapor (H_2O) and can detect NO and methane at specific frequencies.

For example, MODIS (Moderate Resolution Imaging Spectroradiometer) sensors are used for air pollutant distribution mapping by capturing data on aerosol optical depth (AOD) and other atmospheric properties across multiple spectral bands. The AOD data, which indicates the concentration of particulate matter in the atmosphere, is processed and calibrated using ground-based monitoring stations to estimate specific pollutant concentrations, such as $PM_{2.5}$. This information is then visualized through spatio-temporal analysis to identify pollution sources and dispersion patterns, facilitating effective environmental monitoring and management.

Bao et al. (2023) present a significant advancement in retrieval of aerosol optical properties by synergizing multispectral remote sensing data from the Himawari-8 AHI instrument with a Random Forest machine learning algorithm. The methodology, focusing on the utilization of its 16 independent spectral bands, particularly those in the visible and infrared ranges (bands 1–6), specifically incorporates spectral measurements at 0.46 μm, 0.64 μm, and 2.30 μm, along with geometric information such as solar zenith/azimuth and AHI zenith/azimuth angles. This approach allows for the independent retrieval of AOD at red and blue wavelengths, aligning with physical retrieval algorithms and eschewing the need for additional auxiliary data or empirical models. Furthermore, the research emphasizes the extraction of spectral aerosol information through the comparison of independent TOA reflectance between images, diverging from the conventional method of synthesizing data from multiple bands. This technique underscores the significance of utilizing multispectral data to enhance the precision of estimating aerosol optical depth and the Angstrom Exponent. The combination of cutting-edge machine learning algorithms with multispectral remote sensing data represents a significant methodological advancement, enhancing the precision and efficiency of geostationary satellite-based aerosol property retrievals and contributing to the broader field of atmospheric science and climate research. Tao et al. (2023) introduce a multispectral remote sensing technology of multi-angle polarimetric (MAP) measurements, to retrieve detailed aerosol properties. By integrating a physics-informed deep learning (PDL) framework, the study harnesses the advanced deep learning to process complex multispectral data. The PDL method models the intricate relationship between satellite-based MAP observations and aerosol parameters independently, without the need for iterative radiative transfer calculations. This approach not only enhances computational efficiency but also introduces a level of adaptability and speed that significantly benefits operational retrieval processes. The deep learning component, particularly the utilization of Deep Belief Networks (DBNs), plays a pivotal role in establishing accurate and robust predictive models for aerosol optical depth and fine-mode aerosol optical depth, as validated by the study's comparison with AERONET ground-based observations. This research significantly advances the field by showcasing the synergistic potential of multispectral remote sensing and deep learning in atmospheric aerosol characterization, offering a promising avenue for future remote sensing studies.

1.2.2.2 Urban Heat Island

UHI is a phenomenon where urban areas experience significantly higher temperatures than their rural surroundings. This temperature discrepancy arises due to human activities, infrastructure, and land use changes associated with urbanization. Key contributing factors include the extensive use of heat-absorbing materials like concrete and asphalt, reduced vegetation, and anthropogenic heat emissions from vehicles, buildings, and industrial processes. UHI effects can exacerbate energy consumption for cooling, increase air pollution, and negatively impact human health.

Multispectral remote sensing monitors UHI by capturing data across multiple spectral bands to analyze surface temperatures. For the investigation of the -UHI effect, TIR scanning imagery is commonly utilized. TIR remote sensing captures the aggregate radiation from surface objects, the ambient environment, and the atmosphere. In the TIR wavelength range of 10.4 to 12.5 micrometers used in satellite, the contribution of solar radiation is negligible, with most of the detected energy originating from terrestrial emissions. The primary data sources for remote sensing monitoring of the UHI effect predominantly consist of satellite TIR imagery. Commonly utilized satellite data sources include NOAA/AVHRR, Terra, Aqua/MODIS, Terra/ASTER, and Landsat infrared band data.

Extensive studies have been conducted to monitor land surface temperature and UHI by multispectral remote sensing (Zhan et al., 2013). Weng et al.

(2011) innovatively apply multispectral remote sensing, leveraging ASTER imagery, to model UHI and scrutinize their correlation with impervious surfaces and vegetation. Through Linear Spectral Mixture Analysis (LSMA), the research extracts subpixel-level biophysical variables, advancing the characterization of UHI at various spatial scales. Numerous studies aim to leverage multispectral observations to investigate the driving factors of UHI. Li et al. (2023) leverage multispectral remote sensing to analyze UHI effects across global cities, identifying the impact of urban impervious surface growth on UHI intensity and revealing the regulatory roles of vegetation and precipitation. Their study's reconstruction of land surface temperatures from satellite data offers a clear perspective on UHI dynamics, contributing valuable knowledge for urban planning and heat mitigation strategies.

1.3 Conclusion

This chapter has examined the applications of multispectral remote sensing in urban environments, encompassing urban feature extraction and quantitative observations. We have elucidated methodologies for the identification and classification of key urban elements, including green infrastructure, hydrological features, and impervious surfaces, with an emphasis on both knowledge-driven and data-driven approaches. Furthermore, the chapter explored the utilization of multispectral data for quantitative assessments of urban environmental phenomena, particularly atmospheric pollution and UHI effects.

Multispectral remote sensing has emerged as a robust analytical tool for urban studies, offering extensive spatial coverage, temporal consistency, and the capacity to capture data across multiple spectral domains. This technology has facilitated nuanced analyses of urban landscapes, thereby informing urban planning strategies, environmental monitoring protocols, and evidence-based decision-making processes.

The future trajectory of multispectral remote sensing in urban research is promising. Advancements in data fusion techniques, the application of sophisticated machine learning algorithms, improvements in near real-time urban monitoring capabilities, and the synergistic integration with hyperspectral data are poised to enhance its efficacy in urban environmental analysis. However, several challenges persist, including the complexities associated with atmospheric correction, the mixed pixel phenomenon, trade-offs between spatial and temporal resolutions, escalating computational demands for data processing and storage, and the intricacies of validation within heterogeneous urban environments.

Addressing these challenges and capitalizing on emerging opportunities will be instrumental in advancing the application of multispectral remote sensing in urban studies. Future research endeavors should focus on mitigating these limitations and exploring innovative applications to fully leverage the potential of multispectral remote sensing within urban contexts, thereby contributing to more sustainable and resilient urban development practices.

References

Alshehhi, R., Marpu, P. R., Woon, W. L., & Mura, M. D. (2017). Simultaneous extraction of roads and buildings in remote sensing imagery with convolutional neural networks. *ISPRS Journal of Photogrammetry and Remote Sensing, 130*, 139–149. https://doi.org/10.1016/j.isprsjprs.2017.05.002

Bao, F., Huang, K., & Wu, S. (2023). The retrieval of aerosol optical properties based on a random forest machine learning approach: Exploration of geostationary satellite images. *Remote Sensing of Environment, 286*, 113426. https://doi.org/10.1016/j.rse.2022.113426

Li, L., Zhan, W., Hu, L., Chakraborty, T., Wang, Z., Fu, P., Wang, D., Liao, W., Huang, F., Fu, H., Li, J., Liu, Z., Du, H., & Wang, S. (2023). Divergent urbanization-induced impacts on global surface urban heat island trends since 1980s. *Remote Sensing of Environment, 295*, 113650. https://doi.org/10.1016/j.rse.2023.113650

Shi, Q., Liu, M., Marinoni, A., & Liu, X. (2023). UGS-1m: Fine-grained urban green space mapping of 31 major cities in China based on the deep learning framework. *Earth System Science Data, 15*(2), 555–577. https://doi.org/10.5194/essd-15-555-2023

Tao, M., Chen, J., Xu, X., Man, W., Xu, L., Wang, L., Wang, Y., Wang, J., Fan, M., Shahzad, M. I., & Chen, L. (2023). A robust and flexible satellite aerosol retrieval algorithm for multi-angle polarimetric measurements with physics-informed deep learning method. *Remote Sensing of Environment, 297*, 113763. https://doi.org/10.1016/j.rse.2023.113763

Weng, Q., Rajasekar, U., & Hu, X. (2011). Modeling urban heat islands and their relationship with impervious surface and vegetation abundance by using ASTER images. *IEEE Transactions on Geoscience and Remote Sensing, 49*(10), 4080–4089. https://doi.org/10.1109/TGRS.2011.2128874

Yang, X., Qin, Q., Grussenmeyer, P., & Koehl, M. (2018). Urban surface water body detection with suppressed built-up noise based on water indices from sentinel-2 MSI imagery. *Remote Sensing of Environment, 219*, 259–270. https://doi.org/10.1016/j.rse.2018.09.016

Zhan, W., Chen, Y., Zhou, J., Wang, J., Liu, W., Voogt, J., Zhu, X., Quan, J., & Li, J. (2013). Disaggregation of remotely sensed land surface temperature: Literature survey, taxonomy, issues, and caveats. *Remote Sensing of Environment, 131*, 119–139. https://doi.org/10.1016/j.rse.2012.12.014

2

Advances in Nighttime Light Remote Sensing for Urban Observation

Xinyi Shen, Longrui Zhu, Liqun Sun, and Gang Xu

2.1 Introduction

Night time light (NTL) remote sensing is a cutting-edge technology that uses satellite-based observations of artificial illumination to study human activities and socioeconomic dynamics (Elvidge et al., 2017). This technique involves capturing and analyzing the light emitted from the Earth's surface during nighttime hours, providing valuable insights into urban development, economic activity, and human settlement patterns. NTL remote sensing has emerged as a powerful tool because it offers consistent, objective, and spatially explicit data on human presence and activity, often in areas where traditional data collection methods are challenging or inconsistent (Bennett & Smith, 2017; Levin & Zhang, 2017).

The theoretical foundations of NTL remote sensing are rooted in a multidisciplinary approach (Li & Zhou, 2017), incorporating principles from physics, remote sensing, and atmospheric sciences. At its core, this technology relies on detecting and measuring artificial light emitted from the Earth's surface during the nocturnal period, involving complex interactions between electromagnetic radiation and the atmosphere (Kyba et al., 2017). Recent advancements in sensor technology, particularly the introduction of the Visible Infrared Imaging Radiometer Suite (VIIRS) Day/Night Band (DNB), have significantly enhanced the capabilities of NTL remote sensing. These improvements allow for high-sensitivity detection across a wide range of light intensities, enabling the exploration of intra-urban dynamics and revealing disparities in development within cities.

The applications of NTL data are diverse and far-reaching, spanning multiple fields of study and policymaking. Researchers and policymakers use this information to track development trajectories, assess disaster impacts, monitor conflict zones, and estimate Gross Domestic Product (GDP) at sub-national levels (Li et al., 2017). The temporal nature of NTL data allows for the observation of both rapid changes and long-term trends in urban expansion and economic development (Narbudowicz et al., 2016).

 DOI: 10.1201/9781003484219-2

datasets over time. Different sensors, such as the DMSP-OLS and the VIIRS, have been used to capture NTLs. However, these sensors vary significantly in their spatial resolution, spectral sensitivity, and radiometric calibration, leading to discrepancies in the data they produce.

For instance, DMSP-OLS, which was operational from the 1970s until 2013, has a relatively low spatial resolution (about 2.7 km per pixel) and suffers from saturation issues in brightly lit areas. This results in a less detailed and often over-saturated depiction of urban areas. In contrast, the VIIRS sensor, launched in 2011, offers higher spatial resolution (about 750 meters per pixel) and better radiometric sensitivity, allowing for more accurate detection of light variations within cities.

When attempting to analyze long-term trends in urbanization or economic development using both DMSP-OLS and VIIRS data, researchers face the challenge of ensuring that the data from these different sensors are comparable. The differences in resolution and sensitivity can lead to misinterpretations, such as underestimating or overestimating changes in urban growth or economic activity. For example, an area that appears stable in DMSP-OLS data might show significant development when observed with the higher-resolution VIIRS data, leading to conflicting conclusions about urban expansion.

To address cross-sensor consistency issues, researchers have developed various calibration techniques, including statistical methods to harmonize data from different sensors. These techniques involve adjusting the DMSP-OLS data to align with the more accurate VIIRS measurements or using machine learning models to predict what DMSP-OLS data would look like if it had the same characteristics as VIIRS data. However, despite these efforts, achieving perfect consistency remains challenging, particularly when dealing with data collected over several decades.

The challenges of cross-sensor consistency and light pollution underscore the need for careful data processing and interpretation in NTL remote sensing. Both issues have significant implications for applications ranging from urban planning and environmental monitoring to socioeconomic studies, where accurate and consistent data are crucial for informed decision-making.

2.3 Applications of Nighttime Light Remote Sensing

2.3.1 Urban Expansion and Urban Extent

Recent studies have highlighted the utility of NTL time-series data in illuminating important aspects of urbanization. Cities are brightly lit at night, making urban areas easily identifiable in NTL remote sensing data. This capability is vividly demonstrated in Figure 2.3, which depicts the urban expansion of

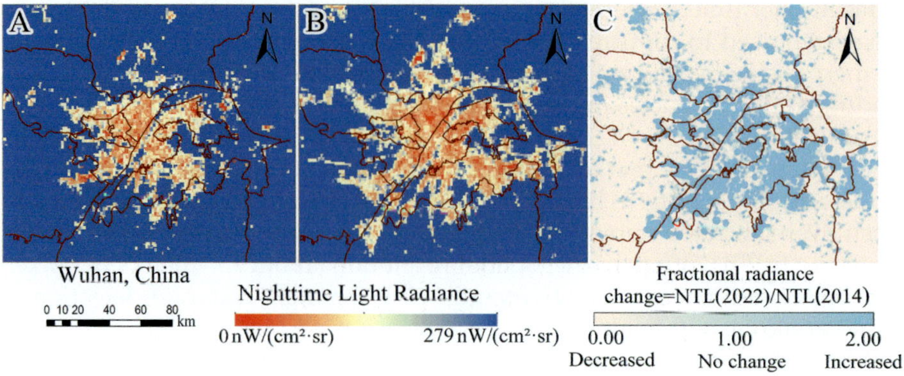

FIGURE 2.3
Observing Urban Expansion in Wuhan, China, Based on Nighttime Light Remote Sensing. Nighttime Light Remote Sensing Imagery from 2014 (A) and 2022 (B) courtesy of NASA's Black Marble. (C) Change from 2014 to 2022.

Wuhan, China, over an eight-year period. The comparative analysis of NTL imagery from 2014 (Figure 2.3A) and 2022 (Figure 2.3B) reveals a marked increase in illuminated areas, indicative of significant urbanization. The spatial distribution of changes (Figure 2.3C) provides compelling evidence of the outward growth of urban development, underscoring the efficacy of NTL data in monitoring urban dynamics.

Urban expansion mapping methods can be categorized into several main types: Threshold-based Methods (L.Imhoff et al., 1997): this method employs a single threshold to map urban areas. Finding an optimal threshold that accurately delineates both large and small cities using NTL data is challenging. A larger threshold effectively identifies large cities but often misses small towns, while a smaller threshold captures small towns but may overestimate the extents of large cities. This issue is further complicated by the overglow effect and variations in lighting standards across countries. As a result, dynamic thresholds are necessary for accurate large-scale and temporally dynamic urban extent mapping. However, thresholds can vary by region and over time, prompting researchers to use multiple thresholds and segmentation-based methods to optimize local thresholds (Zhou et al., 2014).

Machine Learning and Deep Learning Techniques: These methods have gained prominence in geospatial analysis, especially for mapping urban areas using NTL data. They leverage various unstructured data sources, such as NTL images, NTL indices, NDVI, land surface temperature (LST), and point-of-interest densities. For example, Cao et al. combined NTL intensity and NDVI with a region-growing SVM method to identify urban areas in major Chinese cities (Cao et al., 2009). Both supervised and unsupervised machine learning methods have demonstrated higher accuracy compared to traditional thresholding techniques. Incorporating additional features helps mitigate issues like data saturation and overestimation in urban cores, while combining NTL data with Landsat and SPOT images enhances spatial detail and accuracy in complex urban environments.

2.3.2 Economic Growth Prediction

The connection between artificial lighting and urban areas has led many researchers to explore the potential of using nighttime lighting data as an indicator of economic activity. NTL remote sensing data provide a powerful tool for analyzing socioeconomic activities during urbanization and regional development. The strength of incorporating night lights data into economic analyses is therefore in: (1) estimating GDP at finer levels of spatial resolution than are available through official statistics, (2) estimating GDP change (as opposed to levels) at high temporal frequency (Bennie, Davies, Duffy, et al., 2014), and (3) estimating GDP in areas with poor or no reporting (Henderson et al., 2012).

In remote sensing, NTL signals serve as effective indicators of urbanization and regional development, reflecting patterns of population and economic activities. Current research leveraging NTL remote sensing to assess socioeconomic activities can be categorized into four primary areas: (1) examining socioeconomic status at finer spatial scale; (2) tracking socioeconomic dynamics with a higher temporal frequency (Bennie, Davies, Inger, et al., 2014); (3) quantifying socioeconomic parameters in regions with missing data (Henderson et al., 2012); and (4) evaluating development levels to reveal geographic and social issues such as regional inequality, ghost towns, and poverty. By analyzing socioeconomic variables at a grid level, rather than merely relying on administrative units, researchers can gain a more detailed understanding of population and economic changes, facilitating a deeper exploration of these phenomena.

Recent advancements in NTL data, particularly with the newer NPP-VIIRS data, have significantly enhanced the granularity and resolution of socioeconomic research. These improvements allow for detailed mapping of economic activities at various scales, from 5 km to 0.5 km. Understanding the spatial relationships between NTL signals and land surface types is increasingly crucial. Additionally, integrating geolocated big data with traditional statistics and remote sensing products fosters deeper insights into the connections between NTL signals and human activities.

2.3.3 Disaster and Conflict Assessment

Disasters often disrupt electrical services, leading to significant reductions in NTL emissions observable from space. Events such as hurricanes, floods, ice storms, and earthquakes can cause power outages of varying durations, from a few hours to several months, depending on the disaster's severity. NTL monitoring is particularly useful for assessing the impact of major disasters

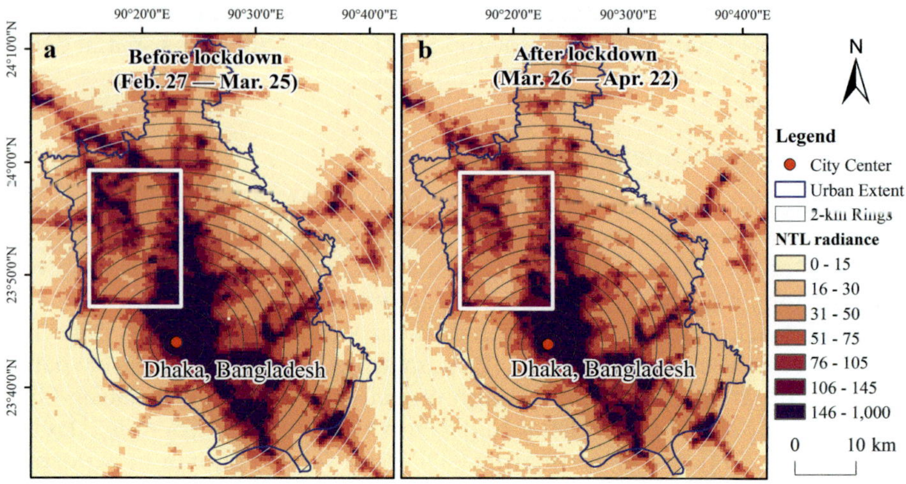

FIGURE 2.4
Spatial distributions of monthly averaged NTL images before (A) and after (B) the lockdown on March 26, 2020 in Dhaka, Bangladesh. (Xu et al., 2021).

and non-weather-related disruptions, where cloud cover might obstruct other observations. Satellites like DMSP/OLS and VIIRS/DNB have proven effective in tracking power outages and recovery. By comparing pre- and post-disaster light levels as elucidated in Figure 2.4, researchers can evaluate damage and restoration efforts, gaining insights into the extent of disruptions and recovery progress.

Human-induced disasters, particularly armed conflicts, have a profound impact on nighttime lighting, offering valuable insights for monitoring these events. Remote sensing data on NTL can reveal crucial information about conflict areas that might be underreported by traditional sources or biased government reports. Research demonstrates that during conflicts, nighttime brightness often declines sharply, reflecting infrastructure damage, potential decreases in population, and unstable power supplies—factors essential for daily life and hospital operations.

For instance, within 36 days of the onset of the Russia-Ukraine conflict, NTL data was employed to assess the total illumination of urban areas in Ukraine. Utilizing NTL ratio indices and dynamic models with time-series data, studies analyzed changes in nighttime illumination on various scales. These analyses revealed that conflicts lead to significant reductions in nighttime lighting, which generally recovers to normal levels post-peace agreements. Countries with frequent conflicts exhibit greater fluctuations in nighttime lighting compared to peaceful nations. NASA's "Black Marble" project has further enhanced monitoring capabilities by providing more frequent nighttime lighting data, improving the temporal resolution for tracking the long-term impacts of these disasters on society and infrastructure.

2.3.4 Environment and Health

NTL remote sensing has emerged as a transformative tool in environmental monitoring, offering a unique vantage point for understanding various environmental phenomena. By leveraging the data derived from NTL observations, researchers can gain insights into patterns and trends that might otherwise be challenging to capture through traditional methods. This approach reveals critical information about human activities and their environmental impacts, thereby enhancing our ability to address pressing environmental issues. These data can be used to monitor the growth of light pollution over time and across different regions (Falchi et al., 2016).

Overall, the integration of NTL data into environmental monitoring offers several advantages. It provides a broad and detailed view of human activities and their environmental impacts, which is often difficult to achieve through other methods. The ability to capture spatial and temporal variations in pollution, light pollution, and heat emissions makes NTL data a powerful tool for researchers and policymakers alike. As advancements in remote sensing technology continue, the application of NTL data in environmental monitoring is likely to become even more sophisticated, further enhancing our capacity to address environmental challenges and promote sustainable development.

Exposure to artificial light at night (ALAN) has been increasingly scrutinized for its potential health impacts, particularly its role in various health conditions such as certain cancers, obesity, and depression. One of the key advancements in assessing the impact of ALAN on health is the use of NTL remote sensing, which leverages satellite imagery to capture and analyze artificial light exposure across different regions. Early studies primarily relied on data from the DMSP. However, a significant development came with the introduction of the VIIRS-DNB. VIIRS-DNB provided a stronger association with its higher spatial resolution compared to DMSP between ALAN and breast cancer (BC) (Rybnikova & Portnov, 2016).

In summary, NTL remote sensing, particularly with advancements such as VIIRS-DNB and International Space Station (ISS) imagery, offers several advantages for assessing the health impacts of artificial light. These technologies allow for precise measurement of light exposure at a higher spatial resolution, improving the accuracy of health impact assessments. The ability to differentiate between light spectra also provides deeper insights into which types of artificial light may be more harmful. As research continues to evolve, the integration of these advanced remote sensing tools will enhance our understanding of how NTL affects human health and inform strategies to mitigate these impacts.

2.4 Recent Progress of Nighttime Light Remote Sensing

2.4.1 High-Resolution Nighttime Light Products

Recent advancements in NTL remote sensing have significantly improved spatial resolution, allowing for more detailed analysis of human activities. The launch of China's SDGSAT-1 in 2021, featuring the Glimmer sensor, offers 100-meter spatial resolution and a 300-kilometer swath width. This marks a significant step forward in mapping urban areas, energy consumption, and economic activities at a finer scale. The integration of SDGSAT-1 data with other high-resolution NTL sources, such as the Jilin-1 constellation. The Jilin-1 Optical A satellite is China's first independently developed high-resolution Earth observation optical imaging satellite. It achieves a ground pixel resolution of 0.72 meters for panchromatic imaging and 2.88 meters for multispectral imaging. The satellite features multiple imaging modes, including conventional push-broom scanning, large-angle side-swing observation, along-track stereo imaging, and multi-strip mosaicing capabilities. The Jilin-1 video satellite, with a ground pixel resolution of 1.12 meters, primarily conducts experimental verification of high-resolution video imaging technology. The Jilin-1 Technology Verification Satellite focuses on validating various multi-mode imaging technologies through experimental verification. Moreover, Luojia 1-01 satellite is the world's first low-orbit micro-nano scientific experimental satellite integrating both remote sensing and navigation capabilities. Initiated by Wuhan University in 2015 and successfully launched on June 2, 2018, it represents a significant advancement in multifunctional satellite technology. The satellite achieves a NTL remote sensing resolution of 130 meters, enabling clear identification of roads and urban blocks, which surpasses both the U.S. DMSP with its OLS sensor (2.7 km resolution) and the VIIRS aboard the NPP satellite (740 meters resolution). This breakthrough facilitates China's transition from conventional surface monitoring to comprehensive socioeconomic development monitoring via remote sensing satellites. Jilin-1 constellation and Luojia 1-01 satellite have enabled a more precise detection of small-scale light sources, including rural settlements and road networks. These improvements contribute to more accurate delineation of urban boundaries and internal structures, advancing our understanding of urbanization processes and socioeconomic disparities at both local and global scales.

High-resolution data not only enhances urban studies but also provides more refined monitoring of human activities in previously underrepresented areas. Smaller light sources, such as isolated rural communities and minor infrastructure, can now be detected and monitored over time. This new capability has opened doors to finer-scale studies of urban sprawl, rural development, and energy use patterns, which were difficult to track using lower-resolution sensors. As satellite capabilities continue to improve, future NTL research will benefit from increasingly detailed and comprehensive datasets, leading to better-informed urban planning and sustainability initiatives.

2.4.2 Spectral and Temporal Advances

Recent innovations in spectral differentiation have overcome the limitations of earlier NTL sensors, such as DMSP-OLS and VIIRS-DNB, which could not distinguish between different lighting types. The development of hyperspectral and multispectral nighttime imaging, such as the DESIS sensor on the International Space Station, allows for the identification of light sources based on their spectral signatures. This ability to differentiate between lighting technologies, such as LED or sodium vapor lamps, offers new insights into urban lighting efficiency and energy consumption. Additionally, spectral differentiation can help assess the ecological impact of various light types on wildlife and natural environments, opening new areas of research in environmental monitoring.

In parallel, the improvement in temporal resolution, led by VIIRS DNB's daily global coverage, has significantly enhanced the monitoring of short-term changes in lighting patterns. This is particularly valuable for studying seasonal variations in human activities, such as fluctuations in energy demand or the impact of events like natural disasters on infrastructure. Combining spectral and temporal advancements, researchers can now perform more dynamic and detailed analyses of socioeconomic and environmental trends, offering new perspectives on energy efficiency, urban growth, and disaster response efforts.

2.5 Conclusions and Outlooks

NTL remote sensing stands at the cusp of a transformative era, driven by technological advancements and interdisciplinary collaborations. The future of this field will likely unfold along three key dimensions: technological innovation, data integration, and application expansion. Technologically, next-generation satellites with enhanced spatial and spectral resolutions will revolutionize nocturnal imagery, enabling more nuanced analyses of urban structures and economic activities. In terms of data integration, the fusion of NTL data with other remote sensing and socioeconomic datasets will provide a more comprehensive understanding of human-environment interactions. This multifaceted approach will be particularly valuable in addressing complex global challenges such as sustainable urbanization and climate change impacts. The application of artificial intelligence and deep learning technologies represents the third frontier, potentially revolutionizing data analysis by uncovering subtle patterns and enabling more efficient processing of vast datasets. However, realizing the full potential of NTL remote sensing will require addressing ongoing challenges in data quality, interpretation, and standardization. As we move forward, collaborative efforts between remote sensing specialists, data scientists, and domain experts will be crucial in developing robust methodologies and ensuring the reliability of NTL datasets across different regions and time periods.

Acknowledgments

This research was supported by the Natural Science Foundation of Hubei Province, China (2023EHA001).

References

Bennett, M. M., & Smith, L. C. (2017). Advances in using multitemporal night-time lights satellite imagery to detect, estimate, and monitor socioeconomic dynamics. Remote Sensing of Environment, 192, 176–197. https://doi.org/10.1016/j.rse.2017.01.005

Bennie, J., Davies, T. W., Duffy, J. P., Inger, R., & Gaston, K. J. (2014). Contrasting trends in light pollution across Europe based on satellite observed night time lights. Scientific Reports, 4(1). https://doi.org/10.1038/srep03789

Bennie, J., Davies, T. W., Inger, R., Gaston, K. J., & Chisholm, R. (2014). Mapping artificial lightscapes for ecological studies. Methods in Ecology and Evolution, 5(6), 534–540. https://doi.org/10.1111/2041-210x.12182

Cao, X., Chen, J., Imura, H., & Higashi, O. (2009). A SVM-based method to extract urban areas from DMSP-OLS and SPOT VGT data. Remote Sensing of Environment, 113(10), 2205–2209. https://doi.org/10.1016/j.rse.2009.06.001

Elvidge, C. D., Baugh, K., Zhizhin, M., Hsu, F. C., & Ghosh, T. (2017). VIIRS night-time lights. International Journal of Remote Sensing, 38(21), 5860–5879. https://doi.org/10.1080/01431161.2017.1342050

Falchi, F., Cinzano, P., Duriscoe, D., Kyba, C. C. M., Elvidge, C. D., Baugh, K., Portnov, B. A., Rybnikova, N. A., & Furgoni, R. (2016). The new world atlas of artificial night sky brightness. Science Advances, 2(6), e1600377. https://doi.org/10.1126/sciadv.1600377

Gang Xu, Tianyu Xiu, Xi Li, Xinlian Liang, Limin Jiao, Lockdown induced night-time light dynamics during the COVID-19 epidemic in global megacities, International Journal of Applied Earth Observation and Geoinformation, Volume 102, 2021, 102421, ISSN 1569-8432, https://doi.org/10.1016/j.jag.2021.102421. (https://www.sciencedirect.com/science/article/pii/S0303243421001288),

Henderson, J. V., Storeygard, A., & Weil, D. N. (2012). Measuring economic growth from outer space. American Economic Review, 102(2), 994–1028. https://doi.org/10.1257/aer.102.2.994

Imhoff, M. L., Lawrence, W. T., Stutzer, D. C., & Elvidge, C. D. (1997). A technique for using composite DMSP/OLS "City Lights" satellite data to map urban area. Remote Sensing of Environment, 61(3), 361–370. https://doi.org/10.1016/s0034-4257(97)00046-1

Kyba, C. C. M., Kuester, T., Sánchez de Miguel, A., Baugh, K., Jechow, A., Hölker, F., Bennie, J., Elvidge, C. D., Gaston, K. J., & Guanter, L. (2017). Artificially lit surface of Earth at night increasing in radiance and extent. Science Advances, 3(11), https://www.science.org/doi/10.1126/sciadv.1701528

Levin, N., & Zhang, Q. (2017). A global analysis of factors controlling VIIRS nighttime light levels from densely populated areas. Remote Sensing of Environment, 190, 366–382. https://doi.org/10.1016/j.rse.2017.01.006

Li, X., Li, D., Xu, H., & Wu, C. (2017). Intercalibration between DMSP/OLS and VIIRS night-time light images to evaluate city light dynamics of Syria's major human settlement during Syrian Civil War. International Journal of Remote Sensing, 38(21), 5934–5951. https://doi.org/10.1080/01431161.2017.1331476

Li, X., & Zhou, Y. (2017). Urban mapping using DMSP/OLS stable night-time light: a review. International Journal of Remote Sensing, 38(21), 6030–6046. https://doi.org/10.1080/01431161.2016.1274451

Narbudowicz, A., Ammann, M. J., & Heberling, D. (2016). Reconfigurable axial ratio in compact GNSS antennas. IEEE Transactions on Antennas and Propagation, 64(10), 4530–4533. https://doi.org/10.1109/TAP.2016.2589975; https://nssdc.gsfc.nasa.gov/nmc/spacecraft/display.action?id=1965-072A

Rybnikova, N. A., & Portnov, B. A. (2016). Outdoor light and breast cancer incidence: a comparative analysis of DMSP and VIIRS-DNB satellite data. InG https://doi.org/10.1016/j.jag.2021.102421

Zhou, Y., Smith, S. J., Elvidge, C. D., Zhao, K., Thomson, A., & Imhoff, M. (2014). A cluster-based method to map urban area from DMSP/OLS nightlights. Remote Sensing of Environment, 147, 173–185. https://doi.org/10.1016/j.rse.2014.03.004

3

Complex Linear Attention Network for Urban Land Cover Mapping from PolSAR Data

Rui Liu and Hongsheng Zhang

3.1 Introduction

The importance of urban area to human is self-explanatory. The dramatic process of urban expansion in recent years has directly altered land cover, with consequent environmental as well as economic impacts. Therefore, timely monitoring of urban cover is not only useful for understanding the process of urbanization and expansion, but also provides research information for environmental and economic studies. Compared with manual field survey, remote sensing can provide large-scale and timely satellite data for urban land cover, and based on the satellite data, expert interpretation or machine learning could be applied to obtain urban land cover types, which greatly saves time and labor costs. Typically, optical remote sensing images are more compatible with the visual presentation of human eye, and their high spectral resolution in the visible and near-infrared band makes them an important data source for urban land cover classification. However, observing over the ground, tropical and subtropical regions tend to have more cloud cover around the year, and urban areas are no exception. Coastal cities inside these regions are also vulnerable to rainfall from the monsoon. These weather conditions would damage the quality of the optical imagery, which is detrimental to urban land cover classification.

Polarimetric synthetic aperture radar (PolSAR) inherits the merits of active imaging mechanism which can be less affected by extreme weather. It's worth noting that the emitted and reflected radiation with different polarization modes are capable of capturing the rich electromagnetic reflection of ground object. These advantages made PolSAR imagery-based urban land cover classification emerged as an alternative fairyland in remote sensing society. As PolSAR possesses unique characteristics, how to extract more discriminative and diverse features for land cover classification has been an enduring research focus. Beginning with single view PolSAR data, the universal feature extraction approaches include polarization scattering matrix decomposition, two-dimensional wavelet transform, polarization fractal features, and

 DOI: 10.1201/9781003484219-3

most recent projection-based method (Han et al., 2023), to name a few. The polarization scattering matrix-based decomposition methods, such as Pauli decomposition and Kennaugh decomposition have emerged as the most popular techniques. Beyond the issue of information loss, various polarization decomposition methods can yield divergent feature representations, resulting in classification outcomes that are highly sensitive to variations in scene characteristics. Consequently, such methods often prove suboptimal for large-scale classification tasks. In recent years, deep learning has exhibited outstanding performance in computer vision and has also demonstrated substantial efficacy within the remote sensing community. In the context of PolSAR image analysis, deep learning methodologies, particularly those grounded in Convolutional Neural Networks (CNNs), have emerged as dominant approaches. For instance, the polarimetric convolutional network proposed by Liu et al. (2019) was designed to extract comprehensive and discriminative features from the polarimetric scattering matrix on a large scale. Drawing upon causal inference theory, Dong et al. (2024) introduced an innovative CNN-based framework for feature enhancement, while Wang et al. (2024) integrated three-dimensional convolution with vision transformer architectures to further improve representation learning. Nevertheless, a fundamental inconsistency arises from the fact that conventional real-valued neural networks are inherently incompatible with the complex-valued nature of polarimetric features, thereby posing significant challenges for the direct application of standard deep learning models to PolSAR data. To mitigate this limitation, researchers have explored complex-valued CNN architectures. Zhang et al. (2017) formulated the essential components and derivative rules within the complex domain, enabling the deployment of complex CNNs for PolSAR data processing. Subsequently, Yu et al. (2023) extended the complex-valued CNN framework to capsule networks to facilitate PolSAR image segmentation. Despite these advances, the intricate semantic structures of land surfaces and their multi-scale variability continue to challenge CNN-based models. Attention mechanisms have therefore been identified as effective strategies to address these issues (Liu et al., 2021). In particular, the squeeze-and-excitation attention mechanism has been incorporated into complex-valued CNNs (Alkhatib et al., 2023), achieving notable performance improvements. However, spatial attention remains computationally demanding, and ensuring the theoretical differentiability of certain operations within the complex domain continues to be a nontrivial challenge.

The challenges motivate us to propose a complex-valued CNN algorithm, equipped with a complex-valued linear attention mechanism for PolSAR data processing, especially towards urban land cover classification task. The major contributions of this chapter are:

1. A complex-valued linear attention mechanism is developed, comprising fully differentiable operations within the complex domain while significantly reducing computational complexity.

2. A classification-oriented complex-valued CNN is designed and integrated with the proposed complex-valued linear attention module to enhance the discriminative representation of PolSAR features.

3.2 Methods

3.2.1 Polarimetric Decomposition Based on Scattering Matrix

Benefiting from the ability to transmit and receive microwaves with different polarizations, the characteristics of the surface can be recorded by PolSAR data in the form of scattering matrix **S**. The microwaves are polarized horizontally (*H*) as well as vertically (*V*), depending on how the transmitted and received microwaves are polarized, **S** can be written as:

$$\boldsymbol{S} = \begin{bmatrix} S_{HH} & S_{HV} \\ S_{VH} & S_{VV} \end{bmatrix} \tag{3.1}$$

Where the first subscript of S is the polarization mode of the transmitting microwave and the other represents the polarization mode of the receiving microwave. Generally further processing of **S** is required, for example, to compute the coherence matrix ***T***, which contains full polarimetric information of the surface, and can be calculated by **S** as:

$$\boldsymbol{T} = \frac{1}{2}\begin{bmatrix} \langle |S_{HH}+S_{VV}|^2\rangle & \langle (S_{HH}+S_{VV})(S_{HH}-S_{VV})^*\rangle & 2\langle (S_{HH}+S_{VV})S_{HV}^*\rangle \\ \langle (S_{HH}-S_{VV})(S_{HH}+S_{VV})^*\rangle & \langle |S_{HH}-S_{VV}|^2\rangle & 2\langle (S_{HH}-S_{VV})S_{HV}^*\rangle \\ 2\langle S_{HV}(S_{HH}+S_{VV})^*\rangle & \langle S_{HV}(S_{HH}-S_{VV})^*\rangle & 4\langle |S_{HV}|^2\rangle \end{bmatrix} \tag{3.2}$$

Obviously, ***T*** is a 3 by 3 matrix, and each element in ***T*** can be denoted as T_{ij}. Because T has symmetry, only six elements of ***T*** are used in the model training, denoted as $T_s = \{T_{11}, T_{12}, T_{13}, T_{22}, T_{23}, T_{33}\}$. It's worth noting that in the selected T_s, there are both real numbers and complex numbers, the real numbers T_{11}, T_{22}, and T_{33} represent surface scattering, double-bounce scattering, and volume scattering, these features provide both unique and consistent attributes of different land cover types. For example, water absorbs most of the microwaves, so the value of these three scatterings is close to zero, whereas vegetation and buildings have strong volume scattering, and some buildings may also present double-bounce scattering. While T_s features exhibit advantages in recognizing certain surface categories, their utility is constrained when dealing with more intricate categories or semantic-related categories, for example, the high-density urban region and the low-density urban region in the coming dataset. This challenge has prompted researchers to try to extract richer features or to find more powerful feature classifiers. Recent deep learning techniques, especially supervised deep learning, are capable of extracting a wide range of features while working as a classifier to output satisfactory classification results in an end-to-end manner. However, two questions remain, one is how real-valued deep learning networks can deal with complex-valued PolSAR data or features, and the other is how to introduce attention mechanisms into complex-valued networks to solve the semantic obfuscation challenge.

3.2.2 Complex-Valued Convolutional Neural Network

There are two key steps to make a CNN complex valued. The first is to define the complex operation for each basic unit in CNN, and the remaining one is to calculate gradient using backpropagation algorithm to update network parameters.

Typically, a CNN is composed of basic convolution, nonlinear activation function, and batch normalization (BN). In the BN layer, there are no parameters that require explicit updating by the gradient, so BN operations in the real domain are equally applicable. For the activation function, it's oriented toward real number and not suitable for complex value. A compromise is to make separate activations for the real and imaginary parts of the complex numbers, despite that it would destroy the valuable phase information in the complex value SAR to some degree. Although it is not backed by a rigorous mathematical theory, this separate activation of real and imaginary parts is an engineering approximation and is widely used in current complex neural networks. The convolution is essentially a superposition of multiplication and summation. If consider the characteristics of SAR data, the CNN should satisfy the phase linearity according to (Zhu et al., 2021), which means for any given phase offset, there is no change in the meaning of the interferogram. However, this theoretical definition is difficult to get a concrete formulaic expression, and this phase linearity property remains corrupted after using the activation function. Currently the input PolSAR can only be considered as complex number, and the convolution of it is transferred from the definition of convolution in the real domain, which can be commonly seen in most complex-valued CNN network. Finally, the fully connected layer is also essentially a combination of multiplication and addition, which is similar to convolution in that it only requires the definition of the real domain to be migrated to the complex domain. When calculating the gradient for these operations, a backpropagation algorithm in the real number domain is often used to calculate the gradient and update the parameters. And in recent years some researchers have pointed out that this approach does not seem to be suitable (Tan et al., 2022) on backpropagation algorithms for complex-valued CNN.

Based on the aforementioned operation, as well as the complex linear attention mechanism that will be mentioned below, we designed the complex linear attention convolutional neural network for PolSAR image classification.

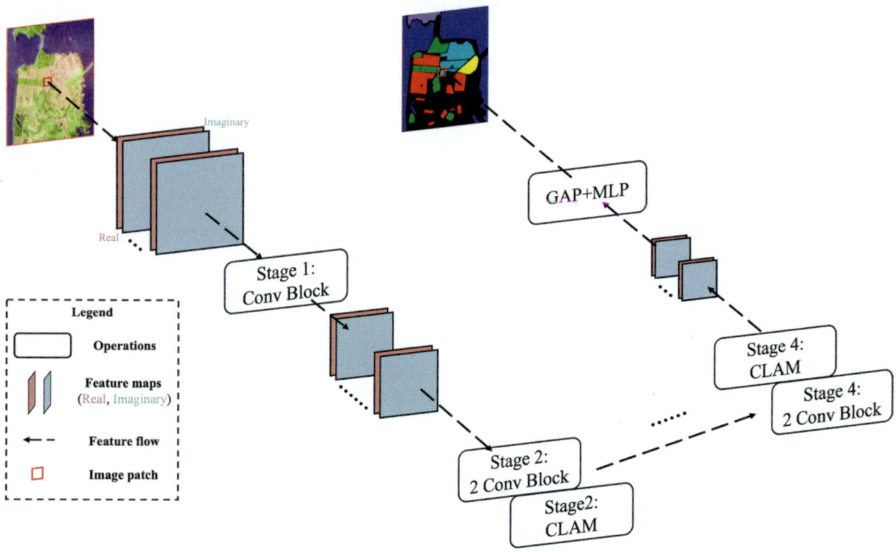

FIGURE 3.1
The complex-valued VGG network with complex-valued linear attention (CLAM) for urban land cover classification with PolSAR.

The overall structure is shown in Figure 3.1. The encoder is based on Visual Geometry Group (VGG) architecture, except the final stage. In the last three stages of the network, there are two convolutional blocks, which are composed of convolution, ReLU and BN layer, and for the first stage, there is one convolution block. For each block, the output channel number is set to be 32, 64, 128, and 128. At the last three stages, after the convolution, the complex linear attention module is added, which is used to capture the global contextual information for the complex-valued PolSAR features. Then the feature maps will be aggregated in the spatial direction, the input and output of each linear attention module remain the form of complex value. Before the final linear layer (fully connected layer), the feature map will be aggregated in the spatial domain and the complex values are converted into norm. Then the fully connected layer will be used as a classifier which takes the norm as input and outputs classification result.

As illustrated in Figure 3.1, for visualization purposes, the input is the image of the Pauli decomposition, but the actual input is PolSAR data in the complex form. With the entire complex-valued VGG network with complex-valued attention mechanism, all the feature maps are purely complex-valued. In general, gradient computation in the complex domain requires the satisfaction of two fundamental conditions: (1) the derivative function must be continuous in the complex domain, and (2) the corresponding operations must fulfill the Cauchy–Riemann equations. As previously discussed, fully connected layers and convolutional operations based on multiplication and addition inherently satisfy the Cauchy–Riemann condition. In contrast, conventional activation functions could not meet the requirement of differentiability in the complex domain. A typical strategy to address this issue is to compute the derivatives of the real and imaginary components separately (Zhang et al., 2017). Supported by these complex-domain operations, each computational unit in the backbone network is designed to be complex-valued. The proposed module adopts a consistent differentiable formulation in the complex-valued linear attention operation, thereby constructing an entirely complex-valued network for PolSAR data processing.

3.2.3 Complex-Valued Linear Attention

The content of this subsection is based on the author's previously published conference paper (Liu et al., 2024). General implementation of the attention mechanism is based on the matrix obtained by multiplying Query (Q) and Key (K), and then apply Softmax function on the multiplication production to generate the attention map. The differentiability of the Softmax function in the complex domain remains a subject of debate, and its use in attention mechanisms significantly increases computational complexity as feature dimensions grow. These challenges motivate the search for alternatives to Softmax that are both compatible with complex-valued operations and computationally efficient. In this work, we define a clear and concise formulation of linear attention within the complex domain, aiming to reduce operation counts while maintaining differentiability and effectiveness. Let Q, K, and V be the Query, Key, and Value matrix with shape of $N \times D$. For a given feature map, N is the number of pixels and D is the number of channels for each pixel. Let $\boldsymbol{q}_i = (q_{i1}, q_{i2}, \ldots, q_{iD})$ be the i_{th} row of Q, and $\boldsymbol{k}_i$, $\boldsymbol{v}_i$ are the same definition. By reviewing the typical attention operation, the first step of calculating attention matrix by Q and K can be written as:

$$QK^T = \begin{bmatrix} \boldsymbol{q}_1 \\ \vdots \\ \boldsymbol{q}_N \end{bmatrix} \begin{bmatrix} \boldsymbol{k}_1^T \cdots \boldsymbol{k}_N^T \end{bmatrix} = \begin{bmatrix} \boldsymbol{q}_1\boldsymbol{k}_1^T & \cdots & \boldsymbol{q}_1\boldsymbol{k}_N^T \\ \vdots & \ddots & \vdots \\ \boldsymbol{q}_N\boldsymbol{k}_1^T & \cdots & \boldsymbol{q}_N\boldsymbol{k}_N^T \end{bmatrix} \tag{3.3}$$

Where $\boldsymbol{k}_i^T$ means the transpose vector of $\boldsymbol{k}_i$, the outcome of Eq. (3.3) is a matrix with shape of $N \times N$. A function f will be applied on the outcome matrix to generate attention matrix A, and then A is multiplied by V, the final output is:

$$Out = f\left(QK^T\right)V = AV \tag{3.4}$$

Let's consider the i_{th} row of the Out, it can be denoted as the matrix multiplication of the i_{th} row of A with V. If V is expanded by row direction, it can be written as:

$$out_i = \boldsymbol{a}_i V = \sum_{j=1}^{N} a_{ij}\boldsymbol{v}_j \tag{3.5}$$

Let a_{ij} denote a scalar element of the attention matrix A. The Softmax function is typically employed as the activation f, but since it requires an $N \times N$ matrix multiplication, the computational complexity scales quadratically with the input size. Following the approaches in (Li et al., 2021; Su, 2020), a Taylor series expansion of the Softmax numerator and denominator can be applied separately in the complex domain, yielding a linearized form of attention. However, this expansion is valid only when the norm $||z||$ is bounded, and the approximation holds over a limited range, making it unsuitable for general complex inputs. To address this, we adopt a linear function f, with slope t and intercept b, to approximate the attention activation. Substituting a_{ij} based on its definition in Eq. (3.3) and applying the linear function f, Eq. (3.5) can then be reformulated as:

$$out_i = \sum_{j=1}^{N}\left(b + t\boldsymbol{q}_i\boldsymbol{k}_j^T\right)\boldsymbol{v}_j = \sum_{j=1}^{N}\left(b\boldsymbol{v}_j + t\boldsymbol{q}_i\boldsymbol{k}_j^T\boldsymbol{v}_j\right) = \sum_{j=1}^{N} b\boldsymbol{v}_j + t\boldsymbol{q}_i\sum_{j=1}^{N}\boldsymbol{k}_j^T\boldsymbol{v}_j \tag{3.6}$$

As shown in Eq. (3.6), the matrix K is first multiplied by V, substantially reducing the computational complexity of the attention operation, particularly when $D \ll N$. In this formulation, the slope of the linear function acts as a scaling factor, while the intercept serves as a residual connection. However, the use of a linear function f may still yield excessively large attention values, even with scaling, which can lead to training instability or non-convergence. To address this, l_2 normalization is applied to q and k vectors, and the cumulants $\boldsymbol{v}_j$ and $\boldsymbol{k}_j^T\boldsymbol{v}_j$ are normalized by dividing by N. enabling the output for each row to be computed efficiently via matrix multiplication.

The architecture of the proposed complex-valued linear attention module is illustrated in Figure 3.2. The module first performs element-wise multiplication between each column of K^T and each row of V, followed by accumulation of the resulting D-dimensional matrices across rows and columns. The accumulated matrix is then multiplied by Q and added to the result of row-wise accumulation of V, completing the attention computation.

3.3 Results and Discussion

3.3.1 Experiment Setup

1. All experiments were performed on a workstation running the Ubuntu operating system and equipped with an NVIDIA GeForce RTX 3090 GPU. For all deep learning-based experiments, the initial learning rate was configured to 0.001, the weight decay was set to 0.02, and the training process was conducted for a maximum of 15 epochs. The batch size was fixed at 128, and all networks were optimized

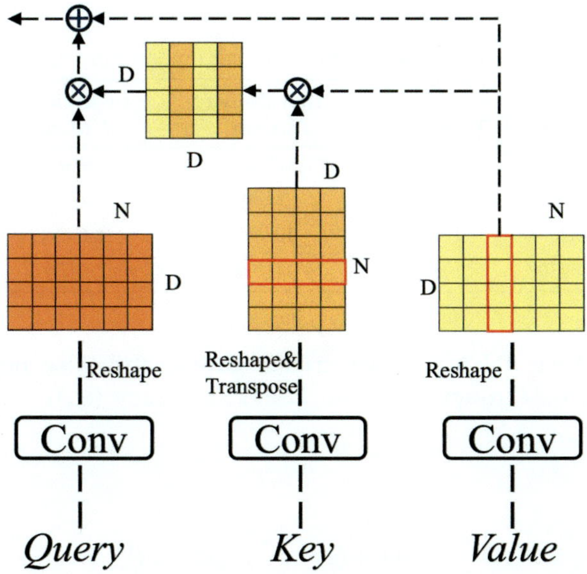

FIGURE 3.2
A basic schematic diagram illustrating the complex-value linear attention mechanism.

using the AdamW optimizer. For the attention parameters *t* and *b*, we set both of them to 1.

2. The experiment dataset is SF-RS2 dataset, where the data region is San Francisco, the data and label are collected and produced by (Liu et al., 2022). The PolSAR data is from Radarsat-2, a C-band SAR satellite owned and operated by a commercial company in Canada. The data collection time is April 2, 2008 with a spatial resolution of 8 meter. For the label, it's manually interpreted to water, vegetation, high-density urban, low-density urban, and developed area, as well as the unlabeled background class on the reference image. In order to accommodate deep learning inputs, the edge region is padded with 16 pixels, to avoid information loss on edge. Then for all deep learning methods, the input patch size is 32. For each category, only 2% samples are selected as training samples. Around 1% of the labeled data are randomly picked as validation data, which is used for all ablation studies. The remaining data served as testing data, which is used to compare performance with different state-of-the-art methods. Detailed information about the split of data can be found in Table 3.1.
3. Evaluation metrics are the commonly used mean Intersection over Union (*mIoU*), overall accuracy (*OA*), and average accuracy (*AA*). For each category, the F_1 score is calculated for comparison. Given a binary classification task, let *TP, FP, TN, FN* be the abbreviates of

TABLE 3.1

Training, Validation, and Testing Sample Number on San Francisco Dataset

Catalog	Water	Vegetation	High-Density Urban	Low-Density Urban	Developed
Training	11238	3905	2136	5467	1285
Validation	5511	1978	1130	2717	662
Testing	536054	192509	107808	266986	62404
Total	552803	198392	111074	275170	64351

true positive, false positive, true negative, and false negative, the Producer's Accuracy (*PA*) and User's Accuracy (*UA*) can be calculated as:

$$PA = \frac{TP}{TP + FN}, \quad UA = \frac{TP}{TP + FP} \tag{3.7}$$

The F_1 score, *OA* and *IoU* are formulated as:

$$F_1 = \frac{2 \cdot PA \cdot UA}{PA + UA} \quad OA = \frac{TP + TN}{TP + TN + FP + FN} \quad IoU = \frac{TP}{TP + FN + FP} \tag{3.8}$$

mIoU and *AA* is the average of *IoU* and *UA* across all categories.

3.3.2 Results on San Francisco Dataset

The study compares the performance of the complex-value linear attention CNN with various algorithms designed for PolSAR image classification, including real-value MLP, complex-value MLP, and real-value CNN based on the VGG structure. The results, presented in Table 3.2, are based on the testing dataset, with model parameters selected from the validation set. The findings reveal that the MLP utilizing single feature points as input struggles with classification accuracy on the San Francisco dataset, as evident in the visual results (Figure 3.3) showing many misclassified pixels. However, incorporating spatial information in the RV-CNN and the proposed method leads to a significant improvement, as demonstrated by the enhanced visual classification results. When looking further at the classified images, we find that for the classification of water bodies, basically all classification methods give better results. This may be due to the fact that water bodies absorb most of the SAR signals, a property that makes them more unique and therefore distinguishable relative to other surface types. In addition, compared to real-valued VGG networks, the proposed method is more advantageous in the recognition of background categories, for example, in the category of

TABLE 3.2

The Numerical Results of Different Methods on San Francisco Dataset, Class Number 0–5 Means Unlabeled, Water, Vegetation, High-Density Urban, Low-Density Urban, and Developed Area

Class	RV-MLP	CV-MLP	RV-VGG	Ours
0	26.45	24.68	83.28	85.42
1	84.50	84.78	98.38	98.16
2	36.84	40.88	94.47	93.76
3	/	1.08	89.89	89.66
4	56.60	57.00	93.75	95.60
5	33.75	37.99	92.16	94.92
OA	56.64	57.74	93.02	93.65
AA	40.84	42.52	93.69	92.27
mIoU	28.46	29.54	85.50	87.06

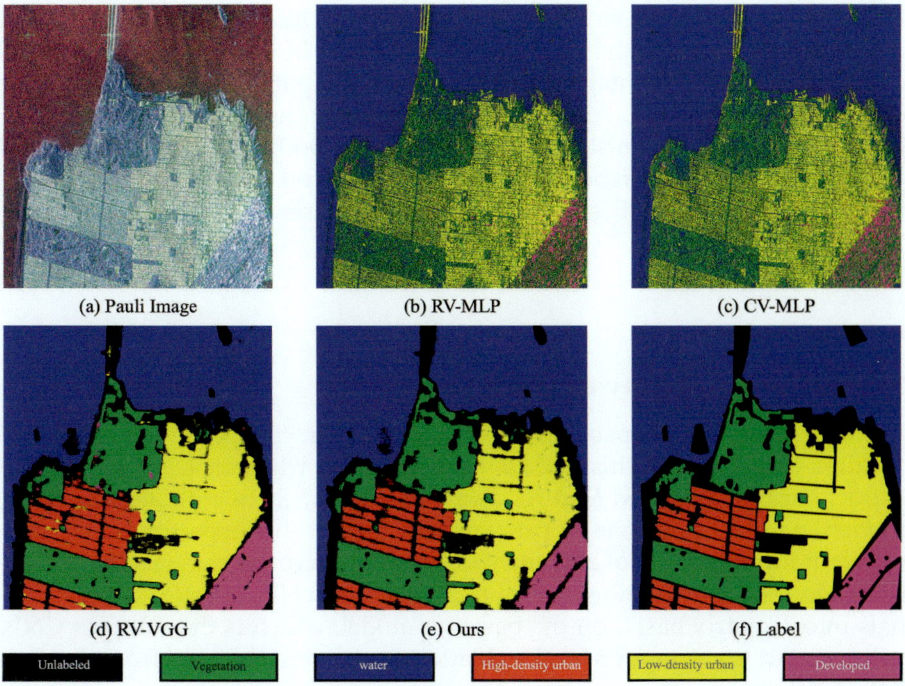

(a) Pauli Image (b) RV-MLP (c) CV-MLP (d) RV-VGG (e) Ours (f) Label (g) Legend

FIGURE 3.3(a–g)
Visual comparison between different classification methods.

vegetation in the middle of the image, the method in this paper can effectively differentiate between the background and vegetation, but the real CNN networks not only have difficulty in differentiating between the background and the vegetation, and even appear to mistakenly categorize vegetation into developed area. This advantage makes our approach visually closer to the reference image in general.

When considered from the numerical results, the observation holds true that our method works best overall, when referred to the *OA* and *mIoU*. Compared to the real-valued CNN networks, although our method lags behind in the categories of water, vegetation, and high-density urban, this lag is very small compared to the performance improvement of our proposed method in the three categories of background, developed area as well as low-density urban. It is also worth noting that for the high-density urban category, the classification results are poor for both real and complex MLPs, and in particular, the real MLPs are not even able to distinguish the category. This may be because the category is closer to low-density urban in terms of pixel-level features and the input of a single pixel point cannot take into account the spatial information within the neighborhoods, thus leading to a failure in the recognition of the category. When an image patch input is used in CNN, the situation becomes better, with better differentiation from the low-density urban and more improvement in the classification accuracy of the high-density urban category. Another advantage of this image patch input is the reduction of *noise* on the classified image. For example, considering the classification of water bodies, there are some misclassification pixels for both real and complex MLP, which not only results in poorer visual perception on the classification image, but also reduces the values of the quantitative metrics.

3.4 Comparison Between Different Input

As deep learning is capable of automatically extracting abstract, high-level features from images, this makes it questionable whether it is necessary to extract various polarized features and then classify them with deep learning models, as has been done in previous studies. In this experiment, we take different inputs as part of ablation study and observe the changes in the classification results by turning the input signals from HH, HV, VV complex signals into T-matrix based on the baseline model, which is the complex CNN network. Accuracy reports for the ablation study were obtained from the validation dataset.

We compare the effects of different types of inputs in terms of both real and imaginary networks. For the real-valued CNN network, we can compare the third and fifth rows of Table 3.3, which have different input but the same real number of convolutional neural networks. There is an increase of the overall

TABLE 3.3

Effects of Different Input Types, the Complex Network, and the Complex Linear Attention Module

Method		Class						Overall Performance		
Input	**Model**	**0**	**1**	**2**	**3**	**4**	**5**	**OA**	**AA**	**mIoU**
HH,HV,VV	RV	80.83	98.35	92.93	91.02	92.44	93.97	92.27	92.37	84.92
HH,HV,VV	CV	83.43	97.39	93.80	90.61	93.57	92.81	92.39	93.18	85.37
T-Matrix	RV	83.41	98.46	94.06	88.95	93.73	92.28	92.91	93.47	85.21
T-Matrix	CV	82.81	97.66	93.28	89.26	94.65	94.24	92.68	91.07	85.51
T-Matrix	CV-CLAM	85.69	98.29	93.65	89.08	95.67	95.47	93.71	92.38	87.17

RV Means Real-Valued CNN, CV Means Complex-Valued CNN, CV-CLAM Is the Proposed Method.

accuracies, albeit to a lesser extent. If we look at the accuracy of each category, it can be seen that the classification accuracy improvement is the highest for vegetative and low-density urban, which corresponds to the advantage of the T-matrix in better describing the scattering characteristics of the surface. When the CNN turns to be complex-valued, the boost from different inputs is smaller. We therefore surmise that the need to extract the coherence matrix from the raw PolSAR data is not particularly relied on when using complex-valued deep learning method.

3.5 The Role of Complex Network and Complex-Valued Linear Attention

First, we compare how much of a classification performance improvement a complex network can deliver with the same input. If the input data are in the HH, HV, and VV bands, comparing the third and fourth rows of Table 3.3, it can be found that when the input data are in complex form, there is a performance improvement in about half of the categories, for example, background, vegetation, low-density urban, etc. For the overall accuracies, the improvement is obvious, especially in *AA*, which gains 1.1%. When the input changes to T-matrix, this performance improvement is still present, as shown in rows five and six of Table 3.3. There are more significant increases in the categories of low-density urban and developed area, and slight improvement in high-density urban. Since the T-matrix is in complex form, it may be because that the complex-valued networks are able to extract the corresponding features from the complex T-matrix in a lossless informative manner as compared to the real-valued networks. Meanwhile, overall accuracy has improved to some extent for both *OA* and *mIoU*.

When referring to the effectiveness of complex linear attention module, we can safely conclude from the sixth and seventh row of Table 3.3 that the attention matrix contributes a lot to the complex-valued CNN. With the exception of the high-density urban category, there is a more significant improvement in classification accuracy for all categories. Of these improvements, the largest performance improvement occurs in the developed area with a 1.23% increase when the background category is not considered, while the smallest performance improvement occurs in the vegetation category. The OA, AA, and mIoU improvements after adding Linear Attention are 1.03%, 1.31%, and 1.66%, all of which are more than 1% improvement.

3.6 Conclusions

This study introduces a novel complex-value linear attention module as the foundation for a pure complex-value CNN designed for processing complex PolSAR data. The module effectively utilizes spatial attention mechanisms to capture global information while also simplifying computational complexity. A notable advantage is its exclusive reliance on complex numbers, eliminating the need for additional processing of complex-value feature maps. The resulting complex neural network inherits the powerful feature representation capabilities synonymous with deep learning methodologies. Through experimentation on the San Francisco dataset, the research findings illustrate remarkable performance enhancements compared to traditional methods such as SVM, RF, and real-value CNNs. These results underscore the promising potential of complex-value neural networks integrated with a sophisticated complex-value attention mechanism for a wide array of applications, signaling a significant advancement in the field of PolSAR (Microwave) data processing and analysis.

References

Alkhatib, M. Q., Al-Saad, M., Aburaed, N., Zitouni, M. S., & Al-Ahmad, H. (2023, July 16–21). PolSAR image classification using attention based shallow to deep convolutional neural network. IGARSS 2023 - 2023 IEEE International Geoscience and Remote Sensing Symposium, Pasadena, California. https://doi.org/10.1109/IGARSS52108.2023.10282338

Dong, H., Si, L., Qiang, W., Zhang, L., Yu, J., Wu, Y., Zheng, C., & Sun, F. (2024). A Novel Causal Inference-Guided Feature Enhancement Framework for PolSAR Image Classification. *IEEE Transactions on Geoscience and Remote Sensing, 62*, 1–16. https://doi.org/10.1109/TGRS.2023.3343380

Han, W., Fu, H., Zhu, J., Zhang, S., Xie, Q., & Hu, J. (2023). A Polarimetric Projection-Based Scattering Characteristics Extraction Tool and Its Application to PolSAR Image Classification. ISPRS Journal of Photogrammetry and Remote Sensing, *202*, 314–333. https://doi.org/10.1016/j.isprsjprs.2023.05.031

Li, R., Zheng, S., Zhang, C., Duan, C., Wang, L., & Atkinson, P. M. (2021). ABCNet: Attentive Bilateral Contextual Network for Efficient Semantic Segmentation of Fine-Resolution Remotely Sensed Imagery. ISPRS Journal of Photogrammetry and Remote Sensing, *181*, 84–98. https://doi.org/10.1016/j.isprsjprs.2021.09.005

Liu, R., Mi, L., & Chen, Z. (2021). AFNet: Adaptive Fusion Network for Remote Sensing Image Semantic Segmentation. IEEE Transactions on Geoscience and Remote Sensing, *59*(9), 7871–7886. https://doi.org/10.1109/TGRS.2020.3034123

Liu, R., Zhang, H., & Lin, J. (2024). PolSAR Image Classification Based on Complex-Value Linear Attention Network. In *2024 IEEE International Conference on Signal, Information and Data Processing (ICSIDP)* (pp. 1–5). IEEE.

Liu, X., Jiao, L., Liu, F., Zhang, D., & Tang, X. (2022). PolSF: PolSAR Image Datasets on San Francisco. In Z. Shi, Y. Jin, & X. Zhang, *Intelligence Science IV Cham*. Springer Nature. https://link.springer.com/chapter/10.1007/978-3-031-14903-0_23

Liu, X., Jiao, L., Tang, X., Sun, Q., & Zhang, D. (2019). Polarimetric Convolutional Network for PolSAR Image Classification. IEEE Transactions on Geoscience and Remote Sensing, *57*(5), 3040–3054. https://doi.org/10.1109/TGRS.2018.2879984

Su, J. (2020). *Exploration of linear attention: Is softmax necessary for attention?* Spaces. Ac.cn. Retrieved on 07/11/2024 from https://kexue.fm/archives/7546

Tan, Z. H., Xie, Y., Jiang, Y., & Zhou, Z. H. (2022). Real-Valued Backpropagation is Unsuitable for Complex-Valued Neural Networks. Advances in Neural Information Processing Systems, *35*, 34052–34063. https://proceedings.neurips.cc/paper_files/paper/2022/hash/dc06d4d2792265fb5454a6092bfd5c6a-Abstract-Conference.html

Wang, L., Gui, R., Hong, H., Hu, J., Ma, L., & Shi, Y. (2024). A 3-D Convolutional Vision Transformer for PolSAR Image Classification and Change Detection. IEEE Journal of Selected Topics in Applied Earth Observations and Remote Sensing, *17*, 11503–11520. https://doi.org/10.1109/JSTARS.2024.3409775

Yu, L., Shao, Q., Guo, Y., Xie, X., Liang, M., & Hong, W. (2023). Complex-Valued U-Net With Capsule Embedded for Semantic Segmentation of PolSAR Image. Remote Sensing, *15*(5), 1371. https://www.mdpi.com/2072-4292/15/5/1371

Zhang, Z., Wang, H., Xu, F., & Jin, Y. Q. (2017). Complex-Valued Convolutional Neural Network and Its Application in Polarimetric SAR Image Classification. IEEE Transactions on Geoscience and Remote Sensing, *55*(12), 7177–7188. https://doi.org/10.1109/TGRS.2017.2743222

Zhu, X. X., Montazeri, S., Ali, M., Hua, Y., Wang, Y., Mou, L., Shi, Y., Xu, F., & Bamler, R. (2021). Deep Learning Meets SAR: Concepts, Models, Pitfalls, and Perspectives. IEEE Geoscience and Remote Sensing Magazine, *9*(4), 143–172. https://doi.org/10.1109/MGRS.2020.3046356

4

Towards 3D Forest Structure Analysis: LiDAR Techniques Advancing Urban Forest Ecosystems

Xiaoling Wang

4.1 The History and Development of LiDAR Remote Sensing

In recent decades, active remote sensing technology has developed rapidly because it can collect data day and night and does not depend on solar radiation. Light detection and ranging (LiDAR), one of the main branches of active remote sensing technology, is widely used in fields such as forestry, agriculture, meteorology, surveying, and archaeology. LiDAR can quantify a target's 3D structural information with high precision by measuring the distance between the sensor and the target and analysing the energy, amplitude, frequency, and phase of the reflected beam (Figure 4.1).

4.1.1 The History of LiDAR

LiDAR, first developed by Hickman and Hogg in 1968, was initially used to measure ocean depths from airborne platforms. Its success led to NASA's creation of Airborne Oceanographic LiDAR (AOL) in the late 1970s, which was used for mapping submarine geomorphology and later applied to land surveys (Arp et al., 1982; Hickman & Hogg, 1969; Hoge et al., 1980). The technology's potential became apparent when it was discovered that LiDAR could penetrate dense vegetation and estimate tree heights. However, it was limited by not being able to detect trees shorter than 8 m (Nelson et al., 1984). Developments in the late 1980s, including global navigation satellite systems (GNSSs) and inertial navigation systems (INSs), allowed for precise real-time scanner location and posture measurements during laser scanning. Professor Ackermann developed the world's first laser cross-section measurement system in 1990, leading to the creation of the first commercial airborne LiDAR, TopScan ALTM 1020, in 1993. LiDAR has since become essential for forest resource surveys and provides key

 DOI: 10.1201/9781003484219-4

a highly accurate remote sensing method for detailed forest plot measurements (Liang and Hyyppä, 2013; Liang et al., 2018), but its high cost and the need for specialized knowledge can be prohibitive. Recent years have seen the introduction of novel techniques for estimating DBH in forests, such as smartphone-based Google Tango and Microsoft Azure Kinect (Fan et al., 2018; Hyyppä et al., 2017; McGlade et al., 2020). In 2020, Apple Inc. launched a LiDAR-based depth sensor in the iPad Pro, augmented with a Reality (AR) application programming interface (API) (Apple Inc., 2020). This technology and various apps offer a new potential alternative remote sensing method. Our aim was to test the feasibility of using the iPad Pro 2020 for estimating urban tree DBH. We focused on: (1) the scanning possibilities of iPad Pro 2020 when urban trees are the objects of scanning; (2) the data acquisition approaches and their impact on DBH estimation accuracy; (3) the impact of different scanning settings on the accuracy of DBH estimation.

We surveyed 101 urban trees with a DBH exceeding 10 cm in Zvolen, Slovakia, with an average DBH of 39.71 ± 19.42 cm. We used the 3DScannerAPP 1.9.1 installed on an iPad Pro 2020 for scanning (Laan Consulting Corp, 2021). This app provides various resolution and confidence settings. We selected "NONE" for masking and used the maximum LiDAR sensor range of 5 m. We experimented with multiple resolutions (10, 15, and 20 mm) and two confidence levels (Low and High). We also tested two scanning modes, single-tree and multiple-tree. Each tree was scanned separately in the single-tree mode, and in the multiple-tree mode, scanning was continuous (Figure 4.4) (Table 4.1). For both modes, we slowly moved around the trunk, keeping the iPad Pro perpendicular to the trunk at breast height and maintaining a distance of 0.5–1 m. We used a circle fitting algorithm to estimate the DBH based on a 2D cross-section of the trunk at 1.3 m height. We initially created an empty 0.5 × 0.5 m grid, normalized Z values for each point, and made a cross-section using points with Z values between 1.28 and 1.32 m. We then grouped these cross-sections based on a maximum distance limit of 10 cm

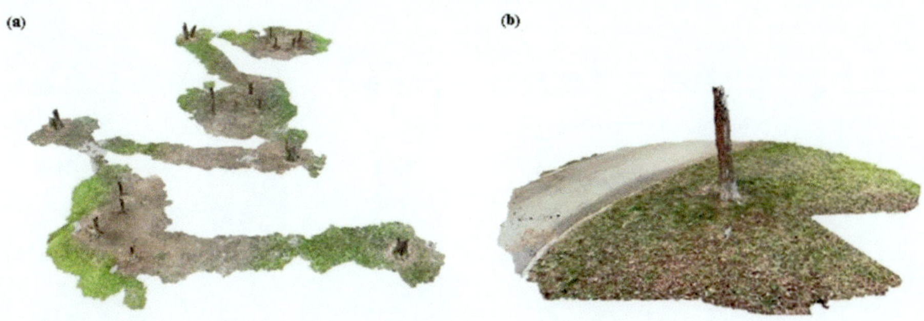

FIGURE 4.4
Demonstration of point clouds of (a) multiple-tree mode with 15 mm resolution and low confidence and (b) single-tree mode with 20 mm resolution and low confidence.

TABLE 4.1

Setting Details

Method	Resolution (mm)	Confidence	Scanning Mode
1	20	High	Multiple
2	20	Low	Multiple
3	15	High	Multiple
4	15	Low	Multiple
5	10	High	Multiple
6	10	Low	Multiple
7	20	High	Single
8	20	Low	Single
9	15	High	Single
10	15	Low	Single
11	10	High	Single
12	10	Low	Single

and a minimum point limit of 200. These groups were visually checked for optimal circle refinement (Koreň et al., 2017). The DBH extraction was conducted using DendroCloud 1.21 software, following the vertical rotation of point clouds in CloudCompare 2.11 software. We synchronized tree positions measured by the Coordinates-GPS Formatter APP with DBH estimates using ArcGIS for Desktop software 10.1. Each estimated DBH layer was overlaid with reference measurements and paired according to the tree positions.

Nearly all methods achieved a 100% detection rate, except method 5, which had a 97% detection rate due to three trees being occluded. These trees, with DBH measurements of 18.97, 20.56, and 32.02 cm, were undetected because of the high resolution and confidence settings. This meant only the best quality points were kept, missing some points at 1.3 m on individual trees (Figure 4.5).

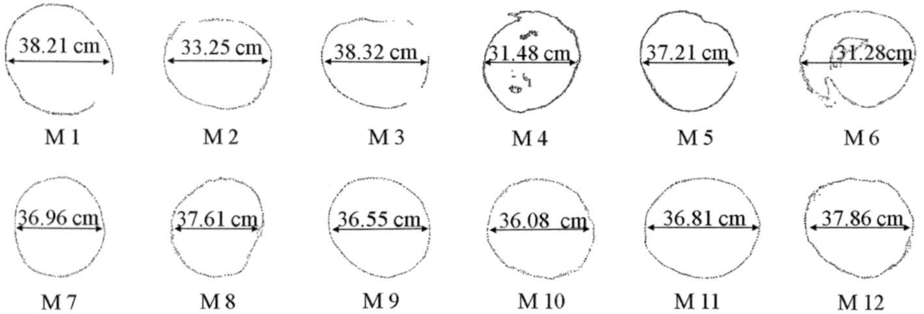

FIGURE 4.5
Illustration of the cross-section at 1.3 m of the same individual tree with DBH = 34.38 cm of multiple-tree mode (Methods 1–6) and single-tree mode (Methods 7–12).

Examples of two 3D models and all point clouds collected are available online (Wang et al., 2021).

The multiple-tree scan mode often required rescanning due to movement between trees, potentially negatively impacting scan quality. Despite this, the multiple-tree mode did not dramatically reduce accuracy compared to the single-tree mode. The most accurate DBH estimation results were obtained using method 8 (rRMSE = 7.00%), method 10 (rRMSE = 7.27%), and method 4 (rRMSE = 7.52%), while method 5 yielded the least consistent results (13.03%). Method 4 presented the best results within the multiple-tree scan mode, whereas method 5 was the least accurate. Method 8 had optimal results among single-tree mode methods, while method 9 had the poorest results. Both modes' most accurate DBH estimations had similar rRMSE (7.52% and 7.00%) (Table 4.2). A boxplot of DBH errors grouped by resolution, confidence, and scanning mode is shown in Figure 4.6. Method 4 (Scanning mode = Multiple-tree mode, Resolution = 15 mm, Confidence = Low) significantly differs from all other methods. Furthermore, the bias of method 4 is the only one that is not substantially different from 0, and at the same time, it is the most accurate multi-tree scanning mode. Method 6 and method 2 significantly underestimated the reference DBH. All methods with single-tree mode are significantly overestimated reference DBH. Also, methods 1, 3, and 5 significantly overestimate reference DBH, under a t-test. Conclusively, we suggested that the iPad Pro 2020 with 3D Scanner APP 1.9.1 (Maximum depth/Range = 5 m, Masking = None, Resolution = 15 mm, Confidence = Low) with multiple-tree mode could provide a feasible, sufficiently accurate, and cost-effective solution for DBH measurements in urban forests.

TABLE 4.2

The Bias and RMSE Resulting from the Estimation of DBH

Method	Bias (cm)	RMSE (cm)	rRMSE (%)
1	2.24	4.18	10.53
2	−2.45	4.13	10.40
3	3.35	4.28	10.78
4	−0.48	2.99	7.52
5	3.26	5.18	13.03
6	−2.36	4.18	10.52
7	3.49	4.23	10.64
8	1.58	2.78	7.00
9	4.02	4.70	11.84
10	1.57	2.89	7.27
11	3.73	4.35	10.94
12	1.51	3.21	8.09

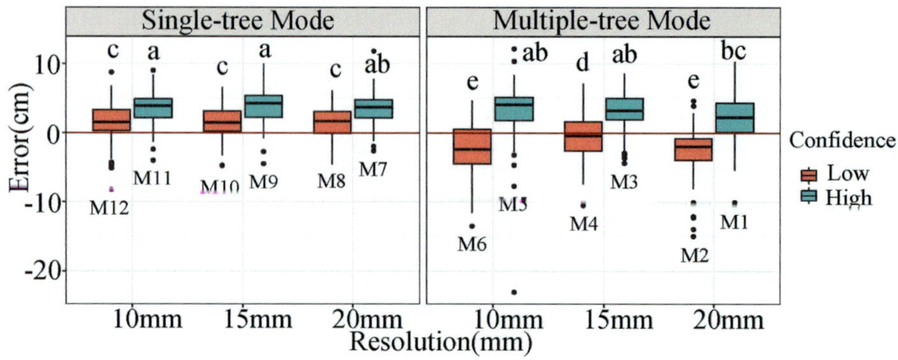

FIGURE 4.6
Boxplot for estimating DBH error using iPad Pro with different settings. A significant difference in analysis among the 12 methods is shown in the graph. Letters a, ab, bc, c, d, and e presented the significance levels (if two variables have different letters, they are significantly different). M is an abbreviation of Method.

4.3.2 The Influence of Vertical Canopy Structure on Cooling Urban Microclimate

Urban forests affect urban microclimates through three key processes: shading, transpiration, and ventilation (Huang et al., 2022; Speak et al., 2020). The urban forest canopy, including its density, horizontal structure, and vertical structure, plays a significant role in these processes. Therefore, assessing how all canopy structural features contribute to cooling and humidifying effects is crucial (Jike et al., 2020). However, capturing vertical structure information has been challenging, limiting understanding of its impact on cooling effects (Kong et al., 2016). Yet, urban forests' cooling and humidifying effect can vary across the day due to factors like solar elevation angle and diurnal wind speed (Armson et al., 2012; Coutts et al., 2016; Huang et al., 2022). Therefore, this study aimed to understand how the canopy structure affects the cooling effect of urban forests at different times of the day using a mobile monitoring system and UAV laser scanner. The novelty of this study lies in the collection and analysis of spatially and temporally extensive air temperature (AT) and absolute humidity (AH) data with high-resolution, complete, and comprehensive canopy structure data. Specifically, we aimed to answer the following questions:

1. How does canopy structure, i.e., canopy density and horizontal and vertical canopy structure, affect the cooling potential of urban forests?
2. Which canopy structural features (i.e., LAI, canopy coverage, mean foliage height, and foliage height diversity) contribute most to the variation in the cooling effect?
3. What is the contribution of each canopy structural feature to the cooling effect at different times of the day?

The research was conducted in Shanghai, a highly populated city in China. The study focused on five urban parks within the city. These parks were chosen due to their similar characteristics, such as distance from the city centre, year of construction, and plant communities. The parks were all part of the Shanghai Green Belt, a critical ecological project in the city (Figure 4.7). A mobile monitoring system was used to record air temperature and humidity. This system was attached to an e-bike, and measurements were taken at fixed routes in each park. The study also utilized two drone scanning systems to gather data on the parks' canopy structures and identify park boundaries, water bodies, and buildings, i.e., UAV laser scanner and multiple-spectrum image systems. The research aimed to understand urban forests' cooling and humidifying effects compared to non-vegetated areas. This was achieved by comparing the air temperature and humidity at each measurement point within the urban forests to those in reference areas. The study also analysed the canopy structures of the urban forests, calculating factors such as leaf area index, canopy coverage, mean foliage height, and foliage height diversity. All measurements were taken during clear, windless days to minimize the influence of weather variables. The research was conducted in the summer of 2021 (Figure 4.8).

The research found significant variations in temperature reduction and humidity increase within the five parks during different times of the day. The highest mean temperature reduction and humidity increase were found between 7:00–10:00 and 11:00–14:00 (Figure 4.9). The study also found a strong relationship between canopy structure and cooling effects but a weaker relationship with humidity increase. Factors such as leaf area index, canopy coverage, mean foliage height, and foliage height diversity all showed varying influences on temperature reduction at different times of the day. In terms of humidity increase, leaf area index and foliage height diversity showed some influence during certain times, but canopy structural features overall had no significant effect on humidity increase during late afternoon and night (Figures 4.10 and 4.11). The study found that vertical canopy structure contributed more than leaf area index and canopy coverage to explain temperature reduction during 11:00–14:00 and 21:00–24:00. However, these canopy structural features contributed less to the variation of daily humidity effects, compared with the cooling effects (Figure 4.12).

Our results illustrated that both an unevenly distributed canopy with low foliage height diversity and an evenly distributed canopy with high foliage height diversity can provide strong cooling effects. Ultimately, we recommended that sites with a high canopy coverage and an unevenly distributed canopy (FHD < 0.3) in high layers are most likely to provide better cooling effects in summer, and tree species such as *Sapindus Saponaria, Koelreuteria bipinnata*, and *Zelkova serrata*, which have wide umbrella canopy shapes, are suggested in Shanghai. Our study shows the contribution of three canopy structural features to the air cooling and humidification of urban forests at different times of the day. This can help understand the cooling mechanisms of urban forests and provide insights for urban forest planning and heat

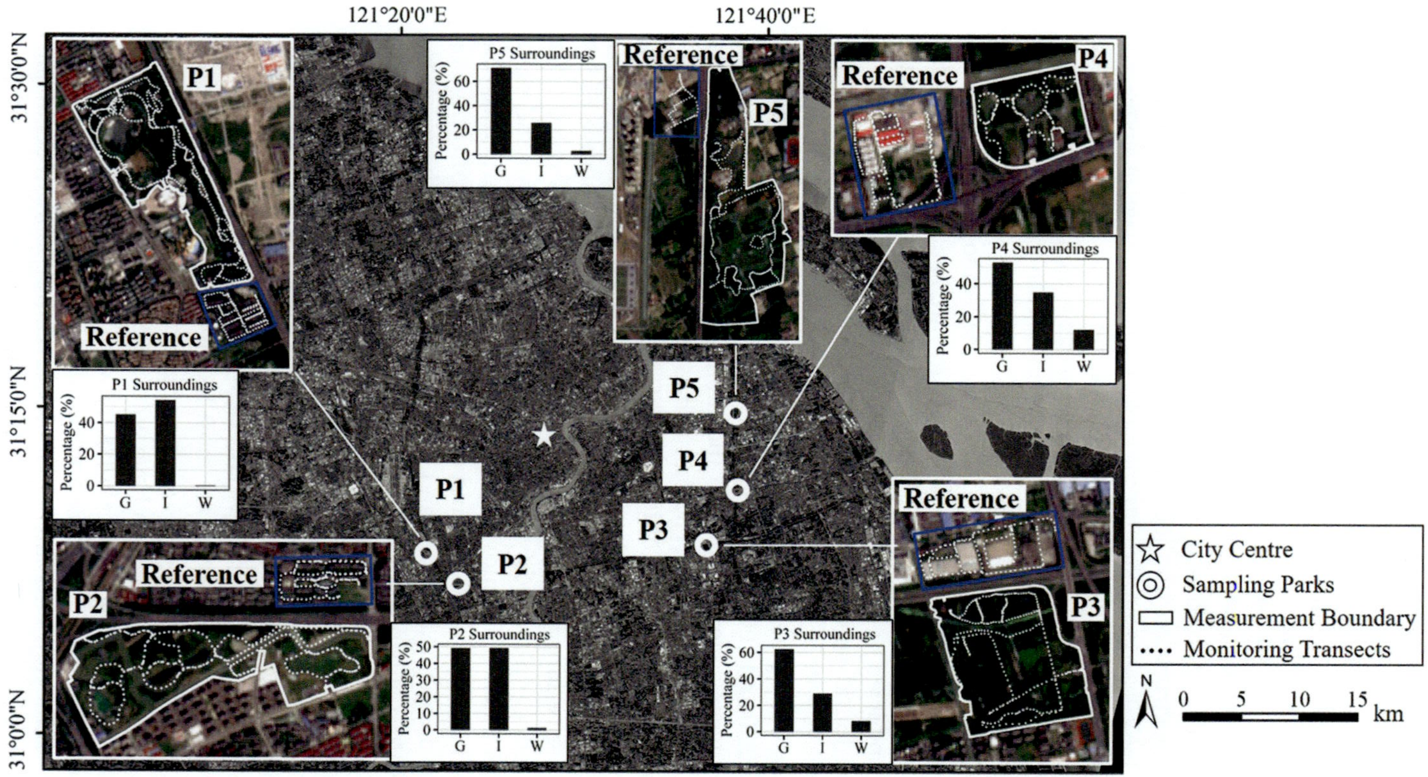

FIGURE 4.7
Location of the five parks (white boundary) and representative reference areas (blue boundary). The temperature measurement points (white points) along the fixed transect in each park (P1–P5) and the reference were recorded every five seconds while measuring. To avoid the advection influence, the cooling or humidifying effects were calculated as the air temperature or absolute humidity anomalies relative to the reference, and the measurement points near the water bodies were excluded. The histograms are the land use percentage of the surrounding background of each park within a 200 m buffer. G, I, and W represent greenspaces, impervious surfaces, and water bodies, respectively.

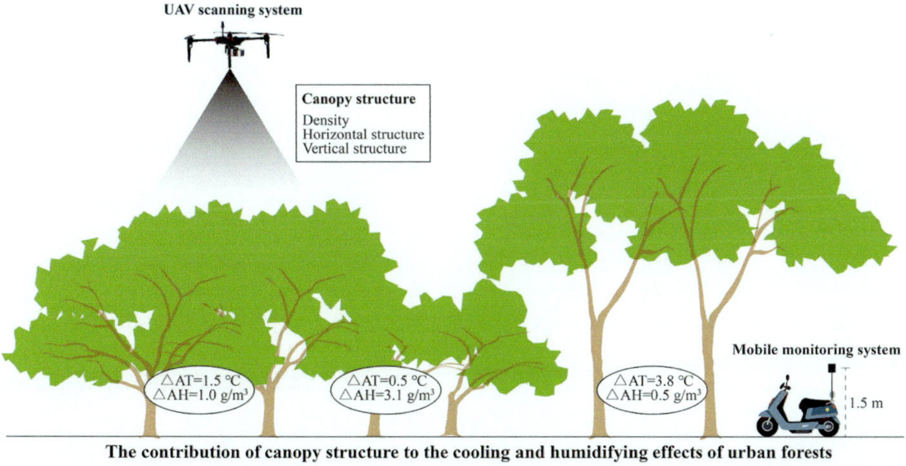

FIGURE 4.8
Conceptual framework for data collection on the cooling effects and canopy structure of urban forests.

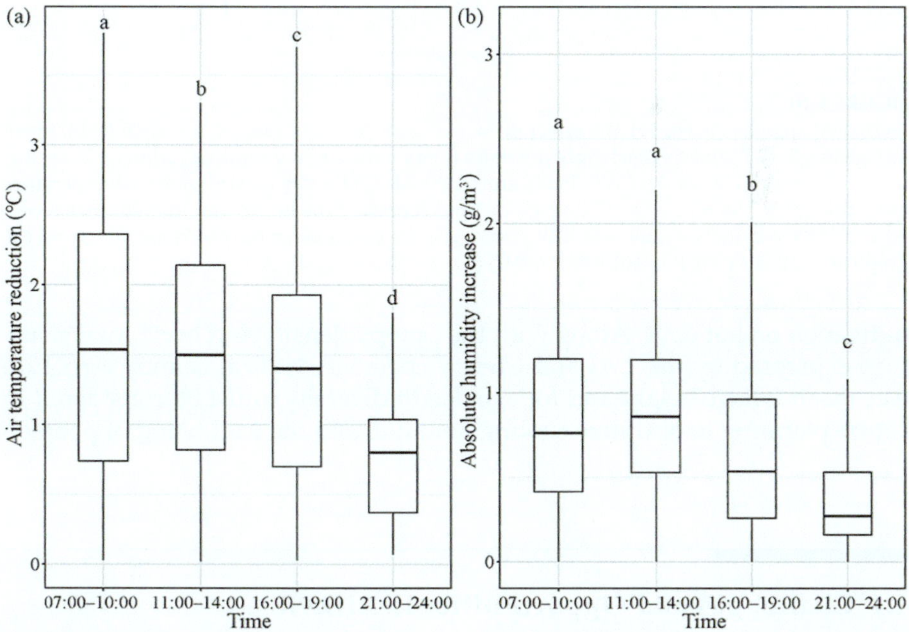

FIGURE 4.9
Boxplots for mean values of (a) air temperature reduction (△AT) and (b) absolute humidity increase (△AH) of five parks during 07:00–10:00, 11:00–14:00, 16:00–19:00, and 21:00–24:00 in summer. Different letters show significant differences among the four times ($P < 0.001$).

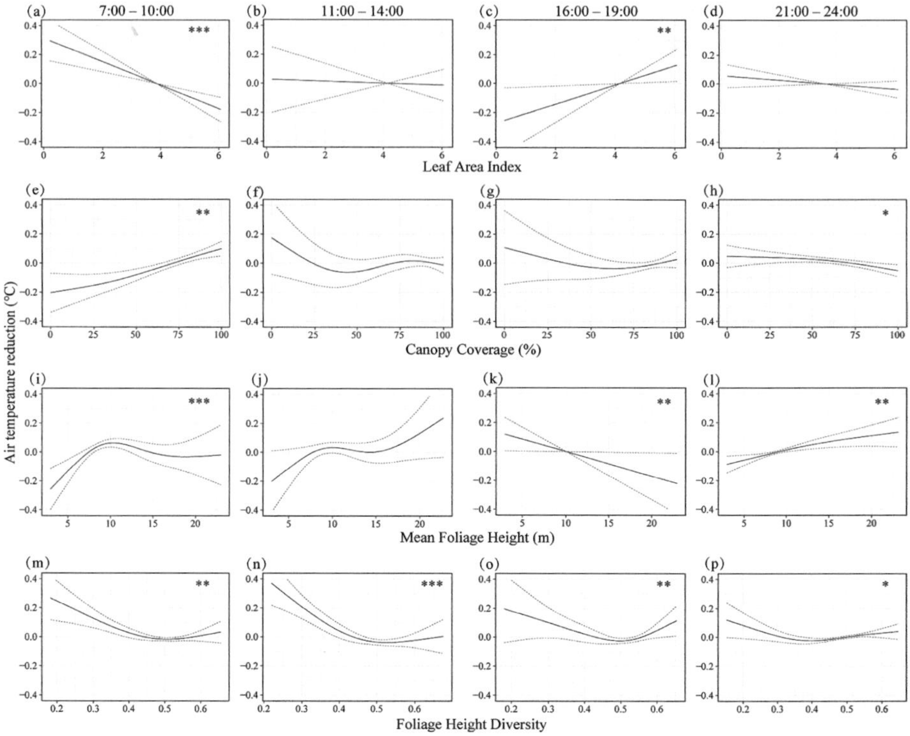

FIGURE 4.10
Estimated smooth curves for the effect of increasing LAI (a–d), canopy coverage (e–h), mean foliage height (i–l), and foliage height diversity (m–p) within a surrounding radius of 20 m during 07:00–10:00, 11:00–14:00, 16:00–19:00, and 21:00–24:00 of a day on summer air temperature reduction (△AT) in Shanghai, China. Black lines represent the mean, and the red lines represent the 95% confidence intervals. The *P values* of the variables in the GAM are shown in the graph at $P < 0.001$ (***), $P < 0.01$ (**), $P < 0.05$ (*).

mitigation on hot days. Along with the canopy density and horizontal structure expressed by the LAI and canopy coverage, vertical canopy structure (i.e., mean foliage height and foliage height diversity) must be considered in future planning to optimize cooling potential and the well-being of people.

4.4 Challenges and Opportunities in LiDAR

Since the 1960s, LiDAR technology has been a game-changer in many areas, including urban forest ecosystems, due to its advantages over traditional measurement methods. The emergence of new technologies like solid-state LiDAR continues to push the boundaries of the field. Despite the challenges

Plowright, A. A., Coops, N. C., Chance, C. M., Sheppard, S. R. J., & Aven, N. W. (2017). Multi-scale analysis of relationship between imperviousness and urban tree height using airborne remote sensing. Remote Sensing of Environment, 194, 391e400.

Reichstein, M., Camps-Valls, G., Stevens, B., Jung, M., Denzler, J., Carvalhais, N., & Prabhat. (2019). Deep learning and process understanding for data-driven Earth system science. Nature, 566(7743), 195e204.

Ren, Z., He, X., Zheng, H., Zhang, D., Yu, X., Shen, G., & Guo, R. (2013). Estimation of the relationship between urban park characteristics and park cool island intensity by remote sensing data and field measurement. Forests, 4(4), 868–886.

Rose, L. S., & Levinson, R. (2013). Analysis of the effect of vegetation on albedo in resi- dential areas: Case studies in suburban Sacramento and Los Angeles, CA. GIScience & Remote Sensing, 50(1), 64e77.

Saatchi, S. S., Harris, N. L., Brown, S., Lefsky, M., Mitchard, E. T., Salas, W., Zutta, B. R., Buermann, W., Lewis, S. L., & Hagen, S. (2011). Benchmark map of forest carbon stocks in tropical regions across three continents. Proceedings of the National Academy of Sciences, 108(24), 9899e9904.

Sankey, T., Donager, J., McVay, J., & Sankey, J. B. (2017). UAV lidar and hyperspectral fusion for forest monitoring in the southwestern USA. Remote Sensing of Environment, 195, 30e43.

Speak, A., Montagnani, L., Wellstein, C., & Zerbe, S. (2020). The influence of tree traits on urban ground surface shade cooling. Landscape and Urban Planning, 197, 103748.

Srinivasan, S., Popescu, S. C., Eriksson, M., Sheridan, R. D., & Ku, N. W. (2014). Multi-temporal terrestrial laser scanning for modeling tree biomass change. Forest Ecology and Management, 318, 304–e317.

Stoker, J. M., Abdullah, Q. A., Nayegandhi, A., & Winehouse, J. (2016). Evaluation of single photon and geiger mode lidar for the 3D elevation program. Remote Sensing, 8(9). Article 767.

Su, Y. J., Guo, Q. H., Xue, B. L., Hu, T. Y., Alvarez, O., Tao, S. L., & Fang, J. Y. (2016). Spatial distribution of forest aboveground biomass in China: Estimation through combination of spaceborne lidar, optical imagery, and forest inventory data. Remote Sensing of Environment, 173, 187e199.

Tian, Y., Zhou, W., Qian, Y., Zheng, Z., & Yan, J. (2019). The effect of urban 2D and 3D morphology on air temperature in residential neighborhoods. Landscape Ecology, 34(5), 1161e1178.

Tooke, T. R., van der Laan, M., & Coops, N. C. (2014). Mapping demand for residential building thermal energy services using airborne LiDAR. Applied Energy, 127, 125e134.

Van Leeuwen, M., Coops, N. C., & Black, T. A. (2015). Using stochastic ray tracing to simulate a dense time series of gross primary productivity. Remote Sensing, 7(12).

Wallace, A. M., McCarthy, A., Nichol, C. J., Ren, X. M., Morak, S., Martinez- Ramirez, D., Woodhouse, I. H., & Buller, G. S. (2014). Design and evaluation of multispectral LiDAR for the recovery of arboreal parameters. IEEE Transactions on Geoscience and Remote Sensing, 52(8), 4942e4954.

Wang, X., Rahman, M. A., Mokroš, M., Rötzer, T., Pattnaik, N., Pang, Y., Zhang, Y., Da, L., & Song, K. (2023). The influence of vertical canopy structure on the cooling and humidifying urban microclimate during hot summer days. Landscape and Urban Planning, 238, 104841.

Wang, X., Singh, A., Pervysheva, Y., Lamatungga, K. E., Murtinová, V., Mukarram, M., Zhu, Q., Song, K., Surový, P., & Mokroš, M. (2021). Evaluation of iPad Pro 2020 LiDAR for Estimating Tree Diameters in Urban Forest 6th International Conference on Smart Data and Smart Cities, Vienna, Austra: ISPRS - International Archives of the Photogrammetry Remote Sensing and Spatial Information Sciences, 105–110.

Sun J, Shi S, Yang J, Gong W, Qiu F, Wang L, et al. (2019). Wavelength selection of the multispectral lidar system for estimating leaf chlorophyll and water contents through the PROSPECT model. Agricultural and Forest Meteorology, 266–267, 43–52.

White, J. D., Ryan, K. C., Key, C. C., & Running, S. W. (1996). Remote sensing of forest fire severity and vegetation recovery. International Journal of Wildland Fire, 6(3), 125e136.

Widlowski, J. L., Mio, C., Disney, M., Adams, J., Andredakis, I., Atzberger, C., et al. (2015). The fourth phase of the radiative transfer model intercomparison (RAMI) exercise: Actual canopy scenarios and conformity testing. Remote Sensing of Environment, 169, 418–437.

Yang, J., & Kang, Z. (2018). Voxel-based extraction of transmission lines from airborne LiDAR point cloud data. IEEE Journal of Selected Topics in Applied Earth Observations and Remote Sensing, 11(10), 3892e3904.

Yi, X., Wang, N., Ren, H., Yu, J., Hu, T., Su, Y., Mi, X., Guo, Q., & Ma, K. (2022). From canopy complementarity to asymmetric competition: The negative relationship between structural diversity and productivity during succession. Journal of Ecology, 110(2), 457–465.

Yu, B., Liu, H., Wu, J., & Lin, W.-M. (2009). Investigating impacts of urban morphology on spatio-temporal variations of solar radiation with airborne LIDAR data and a solar flux model: A case study of downtown Houston. International Journal of Remote Sensing, 30(17), 4359e4385.

Zhao, Q., Myint, S. W., Wentz, E. A., & Fan, C. (2015). Rooftop surface temperature analysis in an urban residential environment. Remote Sensing, 7(9), 12135e12159.

5

Nano-Satellite Constellation

Xiaoxin Zhang

5.1 Overview

In the last two decades, high spatial resolution (<10 m) satellite sensors have been developed and launched by the commercial sector focused predominantly on commercial applications and on gathering geospatial intelligence. The development of commercial low-cost CubeSat constellations provides global coverage data that complement government-funded high-cost moderate (10 to 100 m) to coarse (> 100 m) spatial resolution sensors provided by polar-orbiting satellite series such as Landsat, Sentinel, and Moderate Resolution Imaging Spectroradiometer (MODIS). This chapter provides an in-depth exploration of the transformative potential of nano-satellite technology in urban areas, emphasizing its applications, benefits, and challenges. As cities continue to expand and evolve, the need for timely and accurate data on various urban dynamics has never been greater. Nano-satellites, which are compact, cost-effective satellites capable of capturing HR imagery, present a unique solution to these challenges.

The chapter begins by discussing the architecture and capabilities of nano-satellite constellations, including PlanetScope, Worldview, and Skysat, which offer frequent revisit times and multispectral imaging. These features are particularly beneficial for monitoring urban land use, vegetation cover, and infrastructure development. With daily global coverage and resolutions sufficient to identify individual trees or buildings, nano-satellites provide critical insights that enable urban planners and policymakers to make informed decisions.

A significant focus of the chapter is on the diverse applications of nano-satellite imagery in urban settings. These applications include monitoring urban heat islands, assessing vegetation health, managing water resources, and supporting disaster response efforts. The ability to capture and analyze changes in urban environments over time allows city officials to implement effective greening strategies and infrastructure improvements, ultimately contributing to more sustainable urban living.

Despite the numerous advantages, the chapter also addresses challenges related to the deployment and utilization of nano-satellites, including

DOI: 10.1201/9781003484219-5

data integration, privacy concerns, and the need for robust processing infrastructures.

By highlighting case studies and emerging trends, the chapter underscores the potential of nano-satellite technology to foster resilient urban environments. Ultimately, the insights gained from nano-satellite imagery not only enhance our understanding of urban dynamics but also pave the way for innovative approaches to urban planning and management in an increasingly complex world.

5.2 Multiple Nano-Satellites Constellations

Advancements in satellite technology have led to the development of several widely used satellite imaging systems that provide HR imagery and timely data for various applications. These satellite constellations offer unique capabilities that are essential for monitoring urban areas, managing natural resources, and conducting environmental assessments. Below, we introduce some of the most prominent satellite imaging systems currently in use, highlighting their key features and applications.

5.2.1 RapidEye Collection

RapidEye was a commercial Earth observation satellite constellation operated by Planet Labs and originally developed by the German company RapidEye AG. Launched in August 2008, the RapidEye constellation comprised five identical satellites, each equipped with a multispectral imager that captured data across five spectral bands: blue, green, red, red-edge, and near-infrared (NIR). This extensive spectral range enabled the collection of detailed data suitable for a wide array of applications. Recognized for its efficiency and speed, the RapidEye constellation could revisit the same location on Earth daily, covering over 4 million square kilometers per day with a ground resolution of 5 meters (Table 5.1). This capability was particularly advantageous for industries requiring frequent monitoring and large-scale mapping, including agriculture, forestry, environmental monitoring, and land cover and land use classification.

Beyond agricultural applications, RapidEye's data were instrumental in environmental monitoring efforts, such as tracking deforestation, habitat mapping, and disaster management. By providing clear and consistent imagery, the RapidEye constellation played a crucial role in delivering actionable insights for natural resource management and addressing environmental challenges. With its advanced yet accessible imaging capabilities, RapidEye transformed how industries and governments observed and managed Earth's landscapes, setting a benchmark for future satellite imaging technologies.

TABLE 5.1

Specification of the RapidEye Satellite System

Mission Characteristic	Information
Number of satellites	5
Spacecraft lifetime	7 years
Orbit altitude	630 km in sum-synchronous orbit
Sensor type	Multispectral push-broom imager
Spectral band (nm)	Blue (440–510) Green (520–590) Red (630–685) Red-edge (690–730) Near infra-red (760–850)
Ground sampling distance (nadir)	6.5 m
Spatial resolution (ortho-rectified)	5 m
Swath width	77 km
Onboard data storage	Up to 1500 km of image data per orbit
Revisit time	Daily (off nadir), 5.5 days (at nadir)
Image capture capacity	4 million km^2 per day
Dynamic range	Up to 12 bits

5.2.2 PlanetScope Collection

The PlanetScope mission, initiated by Planet Labs Inc., is an ambitious program developed by the Earth imaging company based in San Francisco (https://docs.planet.com/data/imagery/planetscope/). It employs multiple launches of nano-satellites known as "Doves," constructed using CubeSat 3U models. Each Dove satellite measures approximately 10 × 10 × 30 cm and is equipped with a multi-spectral camera that captures data across four bands in the visible and NIR spectrum, ranging from 455 to 860 nm (Table 5.2). Planet Labs Inc. operates under the mission "To image the entire Earth every day and make global change visible, accessible, and actionable." The company manages a constellation of over 180 CubeSats, capturing multispectral imagery at resolutions between 3.7 and 4.1 meter, depending on altitude. The CubeSat design facilitates a quick build and launch cycle, with an operational lifetime of 2–4 years. While PlanetScope sensors lack onboard calibration devices and strictly maintained orbits, they provide significant opportunities for diverse applications, particularly where high spatial and temporal resolution is required (Roy et al., 2021).

By providing timely and precise multispectral data, PlanetScope enables urban analysts to monitor the dynamics of urban land cover and use in high resolution and make informed decisions, and anticipate urban management and planning (X. Zhang et al., 2025). Its applications extend to disturbance of disaster management, offering critical insights during and after events such as land cover change or flooding, thus facilitating rapid response strategies.

TABLE 5.2
Specification of the RapidEye Satellite System

Mission Characteristic	Information
Number of satellites	430 + (updated on September 2024)
Orbit altitude	475–525 km
Equator Crossing Time	9:30–11:30 am
Sensor type	Multispectral imager
Spectral band (nm)	Coastal blue: 431–452 Blue: 465–515 Green I: 513–549 Green: 547–583 Yellow: 600–620 Red: 650–680 RedEdge: 697–713 NIR: 845–885
Ground sampling distance (nadir)	Dove-C: 3.0–4.1 m Dove-R: 3.0–4.1 m SuperDove: ≥3.7 m
Spatial resolution (ortho-rectified)	3.7–4.1 m, resampled to 3 m
Revisit time	Daily
Image capture capacity	200 million km^2 per day

Ultimately, PlanetScope's cutting-edge satellite technology enhances the understanding of complex urban landscapes and contributes to the development of more resilient and sustainable cities (Neyns et al., 2024; Szabó et al., 2021; X. Zhang et al., 2025). Covering scales from local to regional, PlanetScope imagery have been widely used for HR urban vegetation identification and classification, building classification, and urban tree canopy mapping. The daily recordings from PlanetScope also provide valuable insights with high temporal resolution, essential for monitoring rapid vegetation dynamics, improving detection of critical leaf phenology, and enabling Leaf Area Index estimation through sensor data fusion.

5.2.3 Worldview

The Worldview series of satellites, operated by Maxar Technologies (formerly DigitalGlobe), represents a pinnacle of HR Earth observation technology. This series includes Worldview-1, Worldview-2, Worldview-3, and Worldview-4, all equipped with advanced sensors capable of capturing imagery with a very high spatial resolution of 0.50 meter (https://www.earthdata.nasa.gov/data/tools/worldview). This level of detail enables the discernment of fine features on the Earth's surface, which is essential for various specialized applications (Table 5.3).

In addition to their impressive resolution, the Worldview satellites offer extensive spectral capabilities. Notably, Worldview-3 features eight shortwave

TABLE 5.3

Parameters of the Worldview Satellites Collection

Orbit altitude	**617 km**
Orbit type	Sun-synchronous
Orbit period	97 min
Revisit time	1 day at 1-m GSD resolution 4.5 days at 20° off-nadir (0.59 m GSD)
Equator Crossing Time	10:30 am descending node

infrared (SWIR) bands and 16 multispectral bands, allowing it to capture not only visible light but also to detect materials and conditions that are invisible to the naked eye, such as soil moisture and different types of vegetation.

The Worldview satellite constellation is a vital asset in HR Earth observation, providing critical insights that support informed decision-making across numerous sectors. Its ability to deliver detailed, timely, and reliable data enhances our understanding of the Earth and its landscapes. As technology continues to advance, future iterations of Worldview satellites are expected to further improve their capabilities, potentially providing even more precise data and opening new applications in emerging fields such as AI-driven analytics, climate monitoring, and global environmental protection.

5.2.4 SkySat

The SkySat constellation is an advanced group of HR Earth observation (EO) satellites operated by Planet Labs PBC (https://docs.planet.com/data/imagery/skysat/). Originally developed by Terra Bella and later acquired by Planet Labs from Google in 2017, SkySat satellites are renowned for their capability to capture sub-meter (0.5 to 0.8 meter) resolution multispectral imagery, making them ideal for detailed applications like urban planning, infrastructure monitoring, and disaster response (Jacobsen, 2022).

The first SkySat satellite, SkySat 1, was launched in 2013, with subsequent launches expanding the constellation to improve coverage and frequency (Table 5.4). Notably, three additional satellites, SkySat 19–21, were launched in 2020. The SkySat C generation began with SkySat 3, which was developed as a HR EO model (Murthy et al., 2014). Operating at a nominal altitude of 500 kilometer, each satellite in the constellation offers a revisit period of 4–5 days, facilitating multiple revisits per day collectively. SkySat satellites are equipped with complementary metal oxide semiconductor (CMOS) push-frame cameras, capturing panchromatic and multispectral imagery in blue, green, red, and NIR bands. This setup supports a broad range of Earth observation applications, enhancing the ability to monitor dynamic environments.

SkySat data products are available in four formats: Basic Scene (BS), Ortho Scene (OS), Basemap, and SkySat Ortho Collect (SSC). These products vary

TABLE 5.4

Specification of the SkySat Satellites Collection

Mission Characteristic	Information
Number of satellites	21 (updated on September 2024)
Orbit altitude	SkySat-1–2: 600 km SkySat-3–15: 500 km at launch, lowered to 450 km in early 2020 SkySat-16–18: 400 km
Sensor type	Multispectral imager
Spectral band (nm)	Blue (450–515) Green (515–595) Red (605–695) NIR (740–900) PAN (450–900)
Ground sampling distance (nadir)	[SkySat-1, SkySat-2] -A/B Generation Panchromatic: 0.86 m Multispectral: 1.0 m [SkySat-3–SkySat-15] -C Generation Panchromatic: 0.65 m (0.72 m before 30/06/2020) Multispectral: 0.81 m (1.0 m before 30/06/2020) [SkySat-16–SkySat-21] -C Generation Panchromatic: 0.57 m Multispectral: 0.75 m
Spatial resolution (ortho-rectified)	0.5 m
Swath width	8 km at nadir (SkySat-1–2) 5.9 km at nadir (SkySat-3–15) 5.5 km at nadir (SkySat-16–21)
Onboard data storage	Up to 1500 km of image data per orbit
Revisit time	Constellation: sub-daily, 6–7 times at worldwide average, 12 times max Satellite: 4–5 days (Reference altitude 500 km)
Image capture capacity	4 million km^2 per day

from geometrically raw data with rational polynomial coefficients (RPCs) for advanced processing to geo-corrected formats ready for immediate analysis. The OS product achieves root mean square error (RMSE) accuracy of less than 10 meters, depending on geographical conditions.

The constellation's versatility extends beyond basic imaging, allowing video captures of dynamic changes on Earth and supporting advanced applications through deep learning models, such as those for urban transformation and infrastructure classification. With improvements in satellite agility and tasking latency post-acquisition, SkySat offers rapid response capabilities, vital for timely decision-making in humanitarian and conservation efforts. SkySat's seamless blend of high spatial and temporal resolution empowers a comprehensive understanding of urban landscapes and environmental changes, continuing to make a significant impact across various fields requiring precise Earth observation data.

5.3 Leveraging AI in Nano-Satellite Imagery

In recent years, the rapid development of nano-satellites combined with AI has revolutionized the field of Earth observation, offering unprecedented opportunities for monitoring, analysing, and understanding the change in our earth (Janga et al., 2023). Nano-satellites, often referred to as "CubeSats," are small, lightweight, and cost-effective satellites that have democratized access to space data. When paired with AI, these imagery-rich systems empower diverse applications across environmental monitoring, urban planning, disaster management, and beyond.

Nano-satellites, also known as CubeSats, are small and versatile satellites designed to collect HR imagery at significantly lower costs than traditional satellites (Lucia et al., 2021). The satellites' scalability and capability to form large constellations offer frequent revisits to specific urban areas, ensuring timely and consistent data collection essential for dynamic urban applications. This incessant flow of data becomes a powerful tool when paired with AI, enabling a comprehensive understanding of urban landscapes.

AI, particularly through machine learning and deep learning models, significantly enhances the analysis of imagery data collected by nano-satellites. These AI models excel at processing large datasets, identifying patterns, and providing predictive insights crucial for urban management.

Urban Planning and Infrastructure Monitoring: AI-enabled nano-satellite imagery plays a critical role in mapping and understanding urban growth patterns. By analyzing HR images, AI can classify land use, monitor ongoing construction projects, and detect changes in infrastructure. This information helps urban planners visualize growth trends, optimize land use, and ensure sustainable development. Furthermore, AI can help simulate urban expansion scenarios, allowing cities to proactively plan for future infrastructure needs such as roads, public transport systems, and utilities.

Traffic and Mobility Management: Urban areas face significant challenges in managing traffic congestion and ensuring efficient mobility. AI models analyze satellite imagery to optimize traffic flow, predict congestion points, and facilitate the design of smarter transportation networks. Additionally, AI-driven insights from satellite data can help in planning new public transportation routes and managing parking facilities, making urban mobility more efficient and reducing car emissions.

Environmental Quality and Green Spaces: Nano-satellite imagery, refined by AI, provides detailed assessments of urban green spaces and pollution levels. AI can monitor changes in vegetation cover, helping cities maintain and enhance their parks and green zones. This is crucial for ensuring environmental health and improving the quality of life in cities. By tracking air quality and urban heat islands through multispectral imaging, urban planners can implement strategies to mitigate adverse environmental effects, such as increasing tree cover or creating reflective surfaces to combat urban heat.

Smart City Initiatives: The development of smart cities is fueled by data-driven insights where AI-enhanced satellite imagery plays a key role. These technologies support the integration of Internet of Thing (IoT) devices with urban infrastructures, optimizing resource use and enhancing public services. AI analysis of satellite data aids in energy management, waste collection optimization, and predictive maintenance of city utilities, leading to cost savings and increased service efficiency.

Disaster Preparedness and Response: In densely populated urban environments, timely response to disasters is crucial. Nano-satellites provide real-time data that, when analysed by AI, can assess urban vulnerabilities and forecast potential disaster impacts. This capability enhances preparedness and enables swift responses to events like floods, earthquakes, or fires, ensuring effective evacuation strategies and resource distribution.

5.4 Case Studies for the Application

5.4.1 Urban Land Cover and Use Classification

Land use and land cover (LULC) are crucial factors influencing climate and ecosystem changes through various biophysical and biochemical processes (Hansen et al., 2013). They play a significant role in shaping energy balance, hydrological cycles, carbon dynamics, and ecosystem services, ultimately impacting biodiversity, food security, natural hazards, and public health (Foley et al., 2005). Thus, accurately mapping and monitoring LULC patterns is essential for understanding global environmental changes and facilitating sustainable land management (Hansen et al., 2013).

Historically, urban land use mapping depended on on-site surveys, which were inefficient and geographically limited. The advent of satellite technology since the 1970s has revolutionized this field by providing remote sensing data for multi-scale analysis of urban environments. These data capture the electromagnetic signals from observed objects, allowing effective differentiation of land cover types and widespread generation of LULC maps. Although early satellite systems like Landsat and SPOT provided valuable multispectral images, challenges such as mixed pixels and spectral confusion hindered classification efforts (Chen et al., 2015).

Recent advancements in HR and VHR remote sensing data—such as those from GeoEye, WorldView, and LiDAR—have enhanced the ability to map and classify urban land cover. These technologies enable the differentiation of subpixel information, facilitating more detailed land use classifications. The emergence of social sensing data from smartphones, webcams, and social media also presents new opportunities to incorporate socioeconomic attributes into land use mapping, although the complexity and subjectivity of this data can complicate information extraction (Frias-Martinez & Frias-Martinez, 2014; X. Liu & Long, 2016; Z. Li et al., 2024).

In recent years, deep learning models have gained prominence in land use classification due to their capacity for end-to-end learning and feature extraction. Techniques such as convolutional neural networks (CNNs) and generative adversarial networks (GANs) have demonstrated exceptional performance in remote sensing applications, outperforming traditional models and effectively differentiating complex urban land use categories, especially those with high intraclass variance (Huang et al., 2018; Z. Li et al., 2024).

Despite these advancements, challenges persist in applying deep learning to urban land use classification, including the selection of appropriate data sources, mapping units, and model configurations. Therefore, it is essential to review the successes and limitations of current deep learning applications to guide future research and improve the accuracy of land use identification and urban pattern analysis (Z. Li et al., 2024). Li et al. (2024) utilized PlanetScope imagery and deep learning methods to map urban land cover in Shenzhen (Z. Li et al., 2024). The combination of nano-satellite imagery and advanced deep learning techniques allows for effective classification of urban land use patches.

5.4.2 Mapping Urban Tree Canopy

Urban vegetation are a major component of urban environment, and play a vital role in urban ecosystem, being placed in parks, yards, gardens, and along streets, thus serving as an essential element of urban life (Endreny, 2018; Jackson, 2003). As an essential element of urban vegetation, urban trees offer a range of ecological and socioeconomic benefits, including the mitigation of urban heat islands (Iungman et al., 2023; Schwaab et al., 2021), reduced energy consumption (Ko, 2018), increased carbon sequestration (Nowak & Crane, 2002), air purification (Locosselli et al., 2019), reduced water runoff (B. Zhang et al., 2012), shade (Wang et al., 2020), biodiversity (Alvey, 2006),consequently contributing to improved human well-being and health (Nowak & Crane, 2002). Numerous studies have documented that urban trees provide benefits for municipalities and their residents, and local, regional, and global initiatives have promoted the planting and preservation of urban trees (Eisenman et al., 2021; Yao et al., 2019). Cities present complex and heterogeneous landscapes where vegetation appears in a patchy structure, characterized by irregularly scattered plants with gaps between them. The signals from green grasses, trees, shadows, and buildings are often merged in publicly available satellite images (Shahtahmassebi et al., 2021). Those are characterized by a spatial resolution >10 meter, limiting their applicability to map urban trees, which requires commercial HR datasets (S. Liu et al., 2023). Assessing the services provided by urban greenery requires a suitable scale of analysis, which may vary depending on the service of interest. As an input for studying the urban heat island effect, information on the spatial distribution and density of vegetated areas may be sufficient. However, the services and disservices of urban vegetation can also be studied at a more detailed level, for different species, as is often done for urban trees. As an example, the absorption of airborne pollutants is much larger for some plant species than for others. Consequently, the HR inventory and measurement of urban tree is essential for monitoring and studying the ecological service of urban tree with urbanization and human management.

The emergence of imagery from the PlanetScope nano-satellite constellation, which provides daily global coverage at a resolution sufficient to identify individual trees (approximately 3–5 meters), offers a promising alternative for large-scale mapping and inventory (Roy et al., 2021). X. Zhang et al. (2025) indicated that these images can effectively support the mapping of individual trees at city, and even national scales, as shown in Figure 5.1.

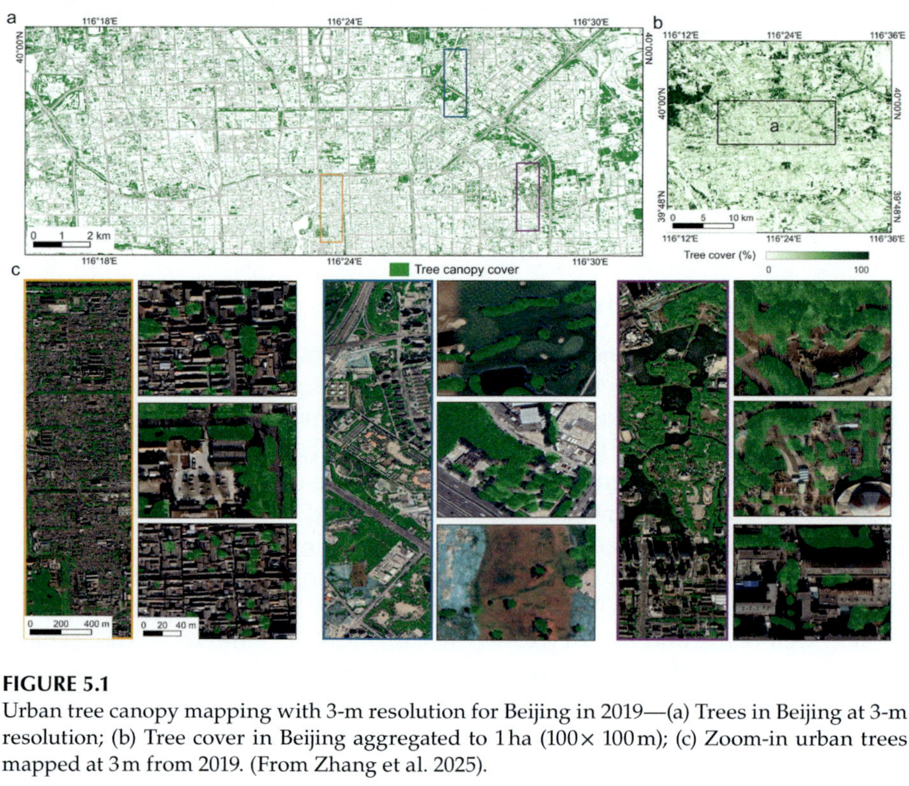

FIGURE 5.1
Urban tree canopy mapping with 3-m resolution for Beijing in 2019—(a) Trees in Beijing at 3-m resolution; (b) Tree cover in Beijing aggregated to 1 ha (100× 100 m); (c) Zoom-in urban trees mapped at 3 m from 2019. (From Zhang et al. 2025).

5.5 Future Opportunities and Challenges

The emergence of nano-satellite constellations has fundamentally transformed the field of earth observation, especially in urban studies. These compact and cost-effective satellites, enhanced with advanced AI technology, provide cities with valuable opportunities to leverage space-based analytics for improved urban planning, management, and policymaking at both local and global scales.

5.5.1 Future Opportunities in Urban Areas

One of the most compelling advantages of nano-satellites is their ability to provide frequent and HR data acquisition (Roy et al., 2021). This capability is essential in urban areas, where rapid changes and development are commonplace. Nano-satellites can revisit specific locations daily, enabling continuous monitoring of urban land cover/use dynamics with frequent human urbanization disturbance. Such capabilities are invaluable for applications like urban ecosystem, sustainable development, where timely information can optimize traffic flow and reduce congestion.

Furthermore, the HR imagery offered by nano-satellites significantly enhances their utility for detailed urban analysis. Equipped with advanced sensors, modern nano-satellites can capture multispectral and even hyperspectral data, enabling urban planners to conduct thorough mapping of city landscapes. This data can effectively identify land-use patterns, assess infrastructure conditions, and detect illegal construction activities. As a result, nano-satellites facilitate precise urban planning initiatives that promote sustainable development through efficient land use and infrastructure deployment.

Another significant advantage of nano-satellites is their cost-effectiveness. Traditional satellite missions are often prohibitively expensive, restricting access primarily to well-funded agencies and corporations. In contrast, nano-satellites democratize access to space by allowing smaller cities and universities to obtain critical data for urban studies. This accessibility fosters innovation and encourages broader participation in urban planning efforts.

In disaster management, nano-satellite imagery is also invaluable (H. Li et al., 2021). Urban areas are vulnerable to various natural and man-made disasters, including floods, earthquakes, and fires (Kapucu, 2012). The rapid data collection capabilities of nano-satellites enable real-time monitoring and assessment of such events, facilitating efficient response and recovery strategies. By providing timely and detailed information, city officials can prioritize actions, coordinate rescue efforts, and allocate resources where they are most needed.

5.5.2 Potential challenges

Despite their potential, the application of nano-satellite imagery in urban studies faces several challenges, particularly in data processing and management. The vast amounts of data generated by these satellites present a significant hurdle; efficiently processing, analyzing, and storing this data necessitates robust computational infrastructure and sophisticated algorithms (Patino & Duque, 2013). For smaller organizations or municipalities with limited technical resources, this can become a substantial bottleneck.

Second, integrating nano-satellite data with existing urban information systems also poses difficulties. Many cities rely on well-established @@Geographic Information Systems (GIS) and other datasets, making it complex to ensure that this new satellite information aligns with existing systems (B. Zhang et al., 2022). Successful integration is crucial for developing comprehensive urban analysis tools that amalgamate multiple data sources, facilitating more informed decision-making.

Additionally, accuracy and validation of satellite data interpretations are additional critical concerns. While HR satellite imagery allows for detailed observations, these insights may not accurately reflect on-the-ground realities without ground trothing (Janssen & Van der Wel, 1994). Obtaining the necessary validation data, particularly in complex urban environments, can be resource-intensive and time-consuming.

Privacy concerns present another significant issue. As imaging capabilities improve, so does the potential for infringing on personal privacy, especially in densely populated urban areas. It is essential to establish protocols and regulations to ensure that the use of nano-satellite data respects individuals' privacy rights.

Finally, technical limitations also remain, primarily regarding the payload capacities of nano-satellites. The interoperability between satellites can be challenging due to high turnover rates stemming from their short lifespans. Furthermore, their limited payload capacity necessitates placing smaller instruments closer together than traditional satellites, which can lead to increased errors between sensors, such as those used in PlanetScope missions. Although technology is advancing, the small size of nano-satellites restricts the diversity and sophistication of the sensors they can carry, limiting the scope of data types available compared to larger, more complex satellite platforms.

In conclusion, while nano-satellites offer tremendous promise for enhancing urban studies through improved data access and analysis, realizing their full potential requires addressing associated challenges, particularly in data management, integration, and ethical considerations. By effectively navigating these hurdles, advanced EO can leverage nano-satellite constellations to foster smarter and more sustainable urban environments.

References

Alvey, A. A. (2006). Promoting and preserving biodiversity in the urban forest. *Urban Forestry & Urban Greening*, *5*(4), 195–201. https://doi.org/10.1016/j.ufug.2006.09.003

Chen, J., Chen, J., Liao, A., Cao, X., Chen, L., Chen, X., He, C., Han, G., Peng, S., Lu, M., Zhang, W., Tong, X., & Mills, J. (2015). Global land cover mapping at 30m resolution: A POK-based operational approach. *ISPRS Journal of Photogrammetry and Remote Sensing, 103*, 7–27. https://doi.org/10.1016/j.isprsjprs.2014.09.002

Eisenman, T. S., Flanders, T., Harper, R. W., Hauer, R. J., & Lieberknecht, K. (2021). Traits of a bloom: A nationwide survey of US urban tree planting initiatives (TPIs). *Urban Forestry & Urban Greening, 61*, 127006.

Endreny, T. A. (2018). Strategically growing the urban forest will improve our world. *Nature Communications, 9*(1), Article 1. https://doi.org/10.1038/s41467-018-03622-0

Foley, J. A., DeFries, R., Asner, G. P., Barford, C., Bonan, G., Carpenter, S. R., Chapin, F. S., Coe, M. T., Daily, G. C., & Gibbs, H. K. (2005). Global consequences of land use. *Science, 309*(5734), 570–574.

Frias-Martinez, V., & Frias-Martinez, E. (2014). Spectral clustering for sensing urban land use using Twitter activity. *Engineering Applications of Artificial Intelligence, 35*, 237–245.

Hansen, M. C., Potapov, P. V., Moore, R., Hancher, M., Turubanova, S. A., Tyukavina, A., Thau, D., Stehman, S. V., Goetz, S. J., Loveland, T. R., Kommareddy, A., Egorov, A., Chini, L., Justice, C. O., & Townshend, J. R. G. (2013). High-resolution global maps of 21st-century forest cover change. *Science, 342*(6160), 850–853. https://doi.org/10.1126/science.1244693

Huang, B., Zhao, B., & Song, Y. (2018). Urban land-use mapping using a deep convolutional neural network with high spatial resolution multispectral remote sensing imagery. *Remote Sensing of Environment, 214*, 73–86.

Iungman, T., Cirach, M., Marando, F., Barboza, E. P., Khomenko, S., Masselot, P., Quijal-Zamorano, M., Mueller, N., Gasparrini, A., Urquiza, J., Heris, M., Thondoo, M., & Nieuwenhuijsen, M. (2023). Cooling cities through urban green infrastructure: A health impact assessment of European cities. *The Lancet, 401*(10376), 577–589. https://doi.org/10.1016/S0140-6736(22)02585-5

Jackson, L. E. (2003). The relationship of urban design to human health and condition. *Landscape and Urban Planning, 64*(4), 191–200. https://doi.org/10.1016/S0169-2046(02)00230-X

Jacobsen, K. (2022). Mapping with skysat images. *The International Archives of the Photogrammetry, Remote Sensing and Spatial Information Sciences, 43*, 99–106.

Janga, B., Asamani, G. P., Sun, Z., & Cristea, N. (2023). A review of practical AI for remote sensing in earth sciences. *Remote Sensing, 15*(16), 4112.

Janssen, L. L., & Van der Wel, F. J. (1994). Accuracy assessment of satellite derived land-cover data: A review. *Photogrammetric Engineering and Remote Sensing, 60*(4), 419–426.

Kapucu, N. (2012). Disaster and emergency management systems in urban areas. *Cities, 29*, S41–S49.

Ko, Y. (2018). Trees and vegetation for residential energy conservation: A critical review for evidence-based urban greening in North America. *Urban Forestry & Urban Greening, 34*, 318–335. https://doi.org/10.1016/j.ufug.2018.07.021

Li, H., Xu, E., & Zhang, H. (2021). High-resolution assessment of urban disaster resilience: A case study of Futian District, Shenzhen, China. *Natural Hazards, 108*(1), 1001–1024.

Li, Z., Chen, B., Wu, S., Su, M., Chen, J. M., & Xu, B. (2024). Deep learning for urban land use category classification: A review and experimental assessment. *Remote Sensing of Environment, 311*. https://doi.org/10.1016/j.rse.2024.114290

Liu, S., Brandt, M., Nord-Larsen, T., Chave, J., Reiner, F., Lang, N., Tong, X., Ciais, P., Igel, C., Pascual, A., Guerra-Hernandez, J., Li, S., Mugabowindekwe, M., Saatchi, S., Yue, Y., Chen, Z., & Fensholt, R. (2023). The overlooked contribution of trees outside forests to tree cover and woody biomass across Europe. *Science Advances, 9*(37). https://doi.org/10.1126/sciadv.adh4097

Liu, X., & Long, Y. (2016). Automated identification and characterization of parcels with OpenStreetMap and points of interest. *Environment and Planning B: Planning and Design, 43*(2), 341–360.

Locosselli, G. M., Camargo, E. P. de, Moreira, T. C. L., Todesco, E., Andrade, M. de F., André, C. D. S. de, André, P. A. de, Singer, J. M., Ferreira, L. S., Saldiva, P. H. N., & Buckeridge, M. S. (2019). The role of air pollution and climate on the growth of urban trees. *Science of The Total Environment, 666*, 652–661. https://doi.org/10.1016/j.scitotenv.2019.02.291

Lucia, B., Denby, B., Manchester, Z., Desai, H., Ruppel, E., & Colin, A. (2021). Computational nanosatellite constellations: Opportunities and challenges. *GetMobile: Mobile Computing and Communications, 25*(1), 16–23.

Murthy, K., Shearn, M., Smiley, B. D., Chau, A. H., Levine, J., & Robinson, M. D. (2014). *SkySat-1: Very high-resolution imagery from a small satellite*. Proceedings volume 9241, 367–378. Sensors, Systems, and Next-Generation Satellites XVIII; 92411E. https://doi.org/10.1117/12.2074163

Neyns, R., Efthymiadis, K., Libin, P., & Canters, F. (2024). Fusion of multi-temporal PlanetScope data and very high-resolution aerial imagery for urban tree species mapping. *Urban Forestry & Urban Greening, 99*, 128410.

Nowak, D. J., & Crane, D. E. (2002). Carbon storage and sequestration by urban trees in the USA. *Environmental Pollution, 116*(3), 381–389. https://doi.org/10.1016/S0269-7491(01)00214-7

Patino, J. E., & Duque, J. C. (2013). A review of regional science applications of satellite remote sensing in urban settings. *Computers, Environment and Urban Systems, 37*, 1–17.

Roy, D. P., Huang, H., Houborg, R., & Martins, V. S. (2021). A global analysis of the temporal availability of PlanetScope high spatial resolution multi-spectral imagery. *Remote Sensing of Environment, 264*. https://doi.org/10.1016/j.rse.2021.112586

Schwaab, J., Meier, R., Mussetti, G., Seneviratne, S., Bürgi, C., & Davin, E. L. (2021). The role of urban trees in reducing land surface temperatures in European cities. *Nature Communications, 12*(1), Article 1. https://doi.org/10.1038/s41467-021-26768-w

Shahtahmassebi, A. R., Li, C., Fan, Y., Wu, Y., lin, Y., Gan, M., Wang, K., Malik, A., & Blackburn, G. A. (2021). Remote sensing of urban green spaces: A review. *Urban Forestry & Urban Greening, 57*. https://doi.org/10.1016/j.ufug.2020.126946

Szabó, L., Abriha, D., Phinzi, K., & Szabó, S. (2021). Urban vegetation classification with high-resolution PlanetScope and SkySat multispectral imagery. *Landscape & Environment, 15*, 66–75. https://doi.org/10.21120/LE/15/1/9

Wang, J., Zhou, W., Jiao, M., Zheng, Z., Ren, T., & Zhang, Q. (2020). Significant effects of ecological context on urban trees' cooling efficiency. *ISPRS Journal of Photogrammetry and Remote Sensing, 159*, 78–89. https://doi.org/10.1016/j.isprsjprs.2019.11.001

Yao, N., Konijnendijk van den Bosch, C. C., Yang, J., Devisscher, T., Wirtz, Z., Jia, L., Duan, J., & Ma, L. (2019). Beijing's 50 million new urban trees: Strategic governance for large-scale urban afforestation. *Urban Forestry & Urban Greening, 44*. https://doi.org/10.1016/j.ufug.2019.126392

Zhang, B., Wu, Y., Zhao, B., Chanussot, J., Hong, D., Yao, J., & Gao, L. (2022). Progress and challenges in intelligent remote sensing satellite systems. *IEEE Journal of Selected Topics in Applied Earth Observations and Remote Sensing*, *15*, 1814–1822.

Zhang, B., Xie, G., Zhang, C., & Zhang, J. (2012). The economic benefits of rainwater-runoff reduction by urban green spaces: A case study in Beijing, China. *Journal of Environmental Management*, *100*, 65–71. https://doi.org/10.1016/j.jenvman.2012.01.015

Zhang, X., Brandt, M., Tong, X., Tong, X., Zhang, W., Reiner, F., Li, S., Tian, F., Yue, Y., Zhou, W., Chen, B., Xiao, X., & Fensholt, R. (2025). A strong but uneven increase in urban tree cover in China over the recent decade. *Nature Cities*, 2(5), 460–469. https://doi.org/10.1038/s44284-025-00227-9

6

The Transformative Role of Social Media Data in Urban Studies

Xiao Huang and Siqin Wang

6.1 Introduction

In the modern urban environment, the pervasive integration of social media platforms has fundamentally transformed the way cities are experienced, interpreted, and governed. Platforms such as Facebook, Twitter, and WeChat, have become essential tools for urban communication, information dissemination, and community engagement. These platforms facilitate real-time connections and interactions among residents, fostering a digital urban community that surpasses traditional physical confines and geographic limitations. The resultant digital footprints from these interactions provide a rich data reservoir, offering valuable insights into urban dynamics, trends, and behaviors.

The influence of social media permeates various aspects of urban life, influencing public opinion (Baum & Potter, 2019), mobilizing community actions (Chen et al., 2021), and acting as a gauge for societal issues and civic participation (Leykin et al., 2018; Liu et al., 2022). In times of urban crises or significant events, social media serves as an essential rapid-response communication channel, enabling the swift dissemination of vital information to diverse urban populations (Wang et al., 2022a and b). This prompt exchange facilitates informed decision-making by both authorities and citizens, thereby enhancing urban resilience and responsiveness. Additionally, social media check-ins and sentiment analysis document public mobility and attitudes, respectively, providing a deeper understanding of urban functionality and public needs (Shen & Karimi, 2016; Yue et al., 2019). The digital traces generated from social media are not merely ephemeral data; they constitute significant analytical points that, when synthesized and examined, disclose complex patterns of urban existence. These patterns yield insights into movement behaviors, congregation points, and prevalent concerns among the populace. By scrutinizing geotagged posts, check-ins, and hashtags, urban planners and researchers can delineate areas of intense activity, detect emergent trends, and decode the sociocultural dynamics of urban spaces.

 DOI: 10.1201/9781003484219-6

In this chapter, we investigate the pivotal role of social media data in urban sensing, demonstrating how these digital traces can be employed to monitor, analyze, and improve urban environments. Through the examination of various case studies and methodologies, we illuminate the transformative impact of social media data across multiple domains that include emergency responses, urban residents' sentiments, place visitation patterns, environmental quality, and participatory planning. We also delve into the limitations of social media data and highlight the critical necessity of integrating digital and physical dimensions of urban spaces to forge smarter, more adaptable cities.

6.2 Social Media's Transformative Potential in Urban Studies

6.2.1 Enhancing Emergency Response

The role of social media platforms in enhancing emergency response capabilities within urban settings is a critical dimension of their transformative impact on urban studies. These platforms facilitate a dual-channel communication pathway that not only enables emergency services to disseminate critical information swiftly but also empowers citizens to report real-time events and emergencies (Wang et al., 2018). This bi-directional exchange of information markedly accelerates the response times of emergency services and enhances situational awareness.

In instances of natural disasters, platforms such as Twitter have been instrumental in facilitating disaster mitigation efforts. Huang et al. (2018a) developed a near real-time flood mapping technique that integrates situational awareness derived from Twitter posts with post-event satellite imagery. Subsequent enhancements include the utilization of advanced convolutional neural network (CNN) architectures for the automatic identification of flood-relevant social media posts (Huang et al., 2019), and the application of sophisticated data fusion frameworks (Huang et al., 2018b). During the 2018 California wildfires and the 2019 Australian wildfires, local authorities and first responders leveraged Twitter to disseminate updates on fire progressions and evacuation orders, providing essential guidance to the public during these chaotic and rapidly evolving crises (Du et al., 2019; Li et al., 2021). Similarly, Facebook's Safety Check feature enables individuals to mark themselves safe during emergencies, delivering immediate reassurance to friends and family while helping to account for persons in affected areas (Lee, 2019).

In essence, the geotagged data from social media posts can be utilized to map emergencies in real-time, identify hotspots, and enable a more targeted response. Analyzing this data facilitates a deeper understanding of the spatial distribution of emergencies, guiding the strategic deployment of resources. The integration of social media data into emergency management systems not only enhances the efficiency and effectiveness of responses but also contributes to a more resilient urban environment. By leveraging the capabilities of these digital platforms, urban areas can better prepare for, respond to, and recover from emergencies, thereby enhancing safety and adaptability in the face of unforeseen challenges. However, this approach necessitates robust data analytics tools and strategies to filter, analyze, and operationalize the vast amounts of data generated during such events, ensuring that the insights derived are actionable and timely.

6.2.2 Gauging Sentiments of Urban Residents

The ability to assess urban residents' sentiments via social media offers a crucial perspective on public mood and opinion. This emotional mapping, interpreted through analytical methods such as natural language processing, provides real-time insights into how urban populations react to local events, policies, or changes within their environments. By analyzing expressions, emotions, and reactions shared on platforms like Twitter, Facebook, and Instagram, urban researchers can gauge a community's pulse, observing fluctuations in sentiment over time or in response to specific situations.

Since the onset of the COVID-19 pandemic, numerous studies have employed various sentiment analysis methods and tools to investigate attitudes and emotions expressed in tweets related to COVID-19 across different geographical regions and scales. A notable example is Wang et al. (2023a), who analyzed 2.1 billion individual-level geotagged tweets posted between February 1, 2020, and March 31, 2021, to monitor and map shifts in negativity across 217 countries and examine its relationship with COVID-19 policies. Their findings reveal significant geographic, demographic, and socioeconomic disparities in negativity across continents, different levels of national income, population density, and COVID-19 infection rates. Similarly, Wang et al. (2022) conducted a nationwide examination of public mental health signals in Australia, highlighting disparities between capital cities and their regional counterparts. They presented a novel empirical framework to systematically measure, classify, and map mental health signals nationwide, thereby assessing the role of public health policies and mental health services in the wake of the global pandemic. Further details on how social media data mining was performed during the COVID-19 pandemic are provided in a comprehensive review article by Huang et al. (2022).

Sentiment analysis in urban settings plays a crucial role in enabling city planners and policymakers to gain a deeper understanding of the diverse needs and concerns of residents, extending well beyond public health issues such as the pandemic. For example, the detection of negative sentiments during urban redevelopment projects can serve as an early indicator of potential resistance or dissatisfaction among the community. This insight allows for proactive engagement with residents and necessary adjustments to project plans to address concerns effectively (Ma et al., 2021; Nik-Bakht & El-Diraby, 2016). Conversely, the identification of positive feedback on social media regarding urban initiatives, such as the introduction of new public transport routes or the renovation of parks, can validate the decisions made by urban planners and provide assurance of public approval (Liang et al., 2022; Lock & Pettit, 2020; Wang et al., 2021). Additionally, in more severe scenarios, sentiment analysis can also help identify distressing trends, such as increased expressions of depression or suicidal thoughts, enabling timely interventions to support mental health (Hamilton et al., 2021; Wang et al., 2023b). By leveraging sentiment analysis, urban planning can become more responsive and adaptive to the evolving dynamics and preferences of urban populations.

As cities continue to evolve, the integration of sentiment analysis with urban planning processes promises to enhance the responsiveness and adaptability of urban spaces, ultimately contributing to more dynamic and empathetic urban environments. This approach not only informs immediate responses but also helps in long-term urban planning and community relationship building.

6.2.3 Understanding Place Visitation Patterns

Understanding place visitation patterns is essential for comprehending the dynamics of urban life, and social media data provides a unique perspective on these patterns. Check-ins and geotagged posts on social media platforms offer a rich dataset that reveals how different demographics utilize various urban spaces. Through the aggregation and analysis of this data, urban planners and researchers can identify popular locations, observe temporal variations in site visitations, and discern the socioeconomic factors influencing urban mobility. Such analysis is crucial for understanding the complex interactions within urban environments and can significantly enhance applications in urban planning.

The strategic use of social year media check-in data can greatly improve urban planning and public service delivery by aligning infrastructure with actual user needs and behaviors. For example, by analyzing check-in frequencies and geographic locations, city planners can optimize public transportation routes to better serve high-traffic areas and manage peak times during major events like concerts, sports and games, or local festivals, thus ensuring efficient crowd management and optimal service availability (Vaingankar et al., 2022; Wang & Huang, 2020). Additionally, this data can guide the strategic placement and timing of public amenities such as parks, libraries, and community centers, as well as temporary facilities like pop-up vaccination clinics or seasonal markets, to effectively meet public demand (Hamstead et al., 2018).

Moreover, the analysis of geotagged posts can also assist in the deployment of emergency services during significant public events or disasters, facilitating a swift and focused response (Reuter et al., 2015). It can further enhance disease modeling by elucidating connectivity patterns among places (Li et al., 2021a 2021,b). Furthermore, examining disparities in the usage of urban spaces revealed by social media check-ins can lead to deeper investigations into accessibility and inclusivity issues (Gomez-Lopez et al., 2017; Hamstead et al., 2018). For instance, lower rates of public space utilization in certain neighborhoods could initiate a comprehensive review of potential barriers such as unsafe pedestrian pathways, inadequate street lighting, or insufficient public transit options. Insights gained from such analyses are invaluable for advocating more equitable urban development, ensuring that improvements in infrastructure and services are inclusively distributed to improve the quality of life across diverse urban communities.

6.2.4 Communicating Environmental Concerns

Social media has transformed the landscape of information dissemination, democratizing access and catalyzing societal discourse on a wide array of subjects. Its impact on fostering environmental consciousness is particularly noteworthy, given the platforms' extensive reach and capability for instantaneous communication. Social media has emerged as a pivotal channel for highlighting environmental issues, articulating challenges, and advocating for sustainable solutions. Various platforms, including Twitter, Weibo, Instagram, Facebook, and TikTok, host a rich array of environmental content. This encompasses detailed discussions on climate change (Williams et al., 2015), biodiversity loss (Havinga et al., 2023), and sustainable practices (Hamid et al., 2017), as well as public mobilizations for recycling initiatives and climate activism (Sujata et al., 2019).

Hashtags such as #ClimateChange, #SaveThePlanet, and #PlasticFree have transitioned from mere online trends to powerful symbols of advocacy and mobilization (Hu et al., 2023). These tags not only enhance awareness and disseminate critical information but also galvanize community action, significantly altering public perceptions and discussions around environmental topics. Social media also serves as a vital platform for environmental activists, including prominent figures like Greta Thunberg. Their compelling narratives and real-time activism not only capture the public's imagination but also achieve widespread distribution, pushing environmental concerns into the limelight. This phenomenon significantly benefits various environmental nongovernmental organizations and movements by providing a dynamic space for interaction, update dissemination, and coordinated action.

Moreover, the role of social media in the realm of science communication cannot be understated, particularly in enhancing public awareness and understanding of environmental sciences (Osterrieder, 2013). Environmental scientists, climatologists, and biologists are increasingly utilizing these platforms to present their findings in accessible and engaging formats, such as infographics, memes, or succinct videos. By adopting these innovative approaches, these professionals are able to spread vital information and counteract the pervasive myths and misinformation prevalent across these digital platforms.

6.2.5 Facilitating Participatory Planning

Participatory planning, facilitated by social media, represents a significant shift in urban development methodologies by empowering citizens to actively engage in the planning process (De Lange & De Waal, 2017). This democratization of planning leverages the capabilities of social media to gather diverse viewpoints and input directly from the community, thus ensuring that urban development is both inclusive and representative of its residents' needs. Social media platforms enable a broad range of stakeholders, including local residents, businesses, and advocacy groups, to voice their opinions, share experiences, and contribute ideas, which are crucial for informed decision-making in urban planning. The immediacy and reach of these platforms allow for real-time feedback and engagement, which is especially useful in dynamic urban environments where conditions and needs can change rapidly.

Utilizing digital platforms, urban planners can effectively engage with communities to solicit feedback on proposed projects or policy changes. This direct interaction is crucial for accurately assessing community sentiments and priorities, enabling planners to make timely adjustments to strategies and enhance residents' sense of ownership over local developments. Bittle et al. (2009) highlight the significant correlation between online engagement and the immediate public realms, such as neighborhoods and local communities. These tools also empower citizens to prioritize issues, foster bipartisan support, and support a hybrid of digital and face-to-face interactions, which enhances collaboration between experts and community members, thereby facilitating local problem-solving. Furthermore, Evans-Cowley and Hollander (2010) examine the benefits of online citizen participation, emphasizing that Information and Communication Technology (ICT) tools significantly improve the public participation process. Bittle et al. (2009) also explore how social media supports e-democracy. During the transition period of the Obama administration, from the election in November 2008 to the inauguration in January 2009, a variety of online tools were tested to engage citizens in a participatory policymaking process. This included a website that featured storytelling, weekly questions, and video messages, with a notable feature that allowed public users to comment on policy announcements and rate others' ideas, thus promoting active public involvement in the decision-making processes that impact them.

Overall, the use of social media in participatory planning is instrumental in enhancing accessibility and inclusivity. Engaging a diverse demographic on these platforms enables city planners to identify and address barriers faced by different community groups, ultimately designing universally accessible urban spaces. Moreover, this approach bolsters urban functionality and promotes equity within the community. Additionally, the integration of social media in participatory planning provides a robust framework for maintaining transparency and accountability. By documenting and sharing the planning process online, municipal governments can sustain public trust and demonstrate their commitment to community-driven decision-making. This transparency not only bolsters public engagement but also motivates active participation in urban development initiatives, allowing residents to see the direct impact of their contributions on municipal planning outcomes.

6.3 A Double-Edged Sword

While social media has proven to be an invaluable resource in urban studies, offering insights into various aspects of urban life and facilitating rapid communication, its limitations and consequences must also be critically assessed. This section explores the challenges associated with using social media data to address the essential topics discussed earlier:

- **Data quality and representativeness**: Social media data are inherently biased towards the demographics of its users, who tend to be younger, more urban, and technologically savvy. This skew can lead to significant gaps in data, particularly from older populations, rural areas, and low-income groups, potentially leading to incomplete or misleading analyses. Such biases can affect urban studies by overemphasizing or underrepresenting certain voices in decision-making processes, particularly in areas like participatory planning and public sentiment analysis.
- **Privacy concerns and ethical use**: The use of social media data raises substantial privacy concerns, as collecting and analyzing data can sometimes occur without the explicit consent of the users. Ethical challenges emerge regarding how data are used, who has access to it, and how much information is revealed about individuals. Mismanagement of this aspect could lead to a breach of trust and reluctance among the public to engage in social media platforms, ultimately hindering their potential as tools for genuine community engagement and emergency communication.
- **Misinformation and data manipulation**: Social media platforms are fertile grounds for the spread of misinformation, which can be particularly detrimental during emergencies or in the mobilization of community actions. The rapid dissemination of false information can lead to panic, misdirected responses, and public distrust in official communications. Urban studies relying on social media data must develop robust methods to identify and filter out misinformation to ensure the reliability of their analyses and the effectiveness of their applications in emergency response and environmental advocacy.

- **Impact on policy and public perception**: While social media can enhance civic engagement and foster democratic participation, its use in urban planning and policymaking must be approached with caution. Over-reliance on data from these platforms can lead to policies that cater more to those who are vocal online, potentially neglecting silent or marginalized groups. Furthermore, the public perception of social media as a tool for surveillance can diminish its utility in participatory planning and community mobilization.

6.4 Conclusion

Social media has emerged as a fundamental element in the contemporary urban fabric, influencing a range of domains from emergency response to environmental activism. Its capacity to generate vast amounts of data that capture the nuances of urban life—often in real time—provides unprecedented opportunities for more nuanced urban sensing and the facilitation of participatory urban planning. Through platforms like Twitter, Facebook, and WeChat, we can now harness these digital traces to gain deeper insights into urban dynamics, enabling more informed and responsive urban governance. However, the utilization of social media data is not without its challenges. Issues of data privacy, representativeness, and the potential for misinformation highlight the need for rigorous standards and ethical guidelines in data handling. Furthermore, the digital divide might skew insights derived from social media, as not all demographic groups engage equally with these platforms. We encourage future research to focus on improving the analytical techniques that filter and interpret social media data, ensuring that it contributes positively to urban development. Embracing both the opportunities and challenges presented by social media will be crucial as we strive to create more adaptable, resilient, and inclusive urban environments. As we continue to navigate this digital era, the interplay between social media data and urban planning will undoubtedly reshape our urban landscapes in profound ways, making the city planning process more dynamic and inclusive than ever before.

References

Baum, M. A., & Potter, P. B. (2019). Media, public opinion, and foreign policy in the age of social media. The Journal of Politics, 81(2), 747–756.

Bittle, S., Haller, C., & Kadlec, A. (2009). Promising practices in online engagement. Occasional Paper, 3.

Chen, Z., Oh, P., & Chen, A. (2021). The role of online media in mobilizing large-scale collective action. Social Media + Society, 7(3), 20563051211033808.

De Lange, M., & De Waal, M. (2017). Owning the City: New Media and Citizen Engagement in Urban Design. In Urban Land Use (pp. 109–130). Apple Academic Press.

Du, H., Nguyen, L., Yang, Z., Abu-Gellban, H., Zhou, X., Xing, W., … Jin, F. (2019, July). Twitter vs news: Concern analysis of the 2018 California wildfire event. In *2019 IEEE 43rd Annual Computer Software and Applications Conference (COMPSAC)* (Vol. 2, pp. 207–212). IEEE.

Evans-Cowley, J., & Hollander, J. (2010). The new generation of public participation: Internet-based participation tools. Planning Practice & Research, 25(3), 397–408.

Gomez-Lopez, I. N., Clarke, P., Hill, A. B., Romero, D. M., Goodspeed, R., Berrocal, V. J., … Veinot, T. C. (2017). Using social media to identify sources of healthy food in urban neighborhoods. Journal of Urban Health, 94(3), 429–436.

Hamid, S., Ijab, M. T., Sulaiman, H., Md. Anwar, R., & Norman, A. A. (2017). Social media for environmental sustainability awareness in higher education. International Journal of Sustainability in Higher Education, 18(4), 474–491.

Hamilton, J. L., Biernesser, C., Moreno, M. A., Porta, G., Hamilton, E., Johnson, K., … Goldstein, T. G. (2021). Social media use and prospective suicidal thoughts and behaviors among adolescents at high risk for suicide. Suicide and Life-Threatening Behavior, 51(6), 1203–1212.

Hamstead, Z. A., Fisher, D., Ilieva, R. T., Wood, S. A., McPhearson, T., & Kremer, P. (2018). Geolocated social media as a rapid indicator of park visitation and equitable park access. Computers, Environment and Urban Systems, 72, 38–50.

Havinga, I., Marcos, D., Bogaart, P., Massimino, D., Hein, L., & Tuia, D. (2023). Social media and deep learning reveal specific cultural preferences for biodiversity. People and Nature, 5(3), 981–998.

Hu, T., Huang, X., & Wang, S. (2023). Actionable Environmental Science Through Social Media Platforms. In Actionable Science of Global Environment Change: From Big Data to Practical Research (pp. 355–371). Cham: Springer International Publishing.

Huang, X., Li, Z., Wang, C., & Ning, H. (2019). Identifying disaster related social media for rapid response: A visual-textual fused CNN architecture. International Journal of Digital Earth 13(9), 1017–1039.

Huang, X., Wang, C., & Li, Z. (2018a). A near real-time flood-mapping approach by integrating social media and post-event satellite imagery. Annals of GIS, 24(2), 113–123.

Huang, X., Wang, C., & Li, Z. (2018b). Reconstructing flood inundation probability by enhancing near real-time imagery with real-time gauges and tweets. IEEE Transactions on Geoscience and Remote Sensing, 56(8), 4691–4701.

Huang, X., Wang, S., Zhang, M., Hu, T., Hohl, A., She, B., … Li, Z. (2022). Social media mining under the COVID-19 context: Progress, challenges, and opportunities. International Journal of Applied Earth Observation and Geoinformation, 113, 102967.

Lee, K. S. (2019, October). Explicit disaster response features in social media: safety check and community help usage on Facebook during Typhoon Mangkhut. In *Proceedings of the 21st International Conference on Human-Computer Interaction with Mobile Devices and Services* (pp. 1–12).

Leykin, D., Lahad, M., & Aharonson-Daniel, L. (2018). Gauging urban resilience from social media. International Journal of Disaster Risk Reduction, 31, 393–402.

Li, L., Ma, Z., & Cao, T. (2021). Data-driven investigations of using social media to aid evacuations amid Western United States wildfire season. Fire Safety Journal, 126, 103480.

Li, Z., Huang, X., Hu, T., Ning, H., Ye, X., Huang, B., & Li, X. (2021b). ODT FLOW: Extracting, analyzing, and sharing multi-source multi-scale human mobility. PLoS One, 16(8), e0255259.

Li, Z., Huang, X., Ye, X., Jiang, Y., Martin, Y., Ning, H., … Li, X. (2021a). Measuring global multi-scale place connectivity using geotagged social media data. Scientific Reports, 11(1), 14694.

Liang, H., Yan, Q., Yan, Y., Zhang, L., & Zhang, Q. (2022). Spatiotemporal study of park sentiments at metropolitan scale using multiple social media data. Land, 11(9), 1497.

Liu, D., Feng, X. L., Ahmed, F., Shahid, M., & Guo, J. (2022). Detecting and measuring depression on social media using a machine learning approach: Systematic review. JMIR Mental Health, 9(3), e27244.

Lock, O., & Pettit, C. (2020). Social media as passive geo-participation in transportation planning–how effective are topic modeling & sentiment analysis in comparison with citizen surveys? Geo-spatial Information Science, 23(4), 275–292.

Ma, Y., Yang, Y., & Jiao, H. (2021). Exploring the impact of urban built environment on public emotions based on social media data: A case study of Wuhan. Land, 10(9), 986.

Nik-Bakht, M., & El-Diraby, T. E. (2016). Sus-tweet-ability: Exposing public community′ s perspective on sustainability of urban infrastructure through online social media. International Journal of Human-Computer Studies, 89, 54–72.

Osterrieder, A. (2013). The value and use of social media as communication tool in the plant sciences. Plant Methods, 9, 1–6.

Reuter, C., Ludwig, T., Friberg, T., Pratzler-Wanczura, S., & Gizikis, A. (2015). Social media and emergency services?: Interview study on current and potential use in 7 European countries Interview study on current and potential use in 7 European countries. International Journal of Information Systems for Crisis Response and Management (IJISCRAM), 7(2), 36–58.

Shen, Y., & Karimi, K. (2016). Urban function connectivity: Characterisation of functional urban streets with social media check-in data. Cities, 55, 9–21.

Sujata, M., Khor, K. S., Ramayah, T., & Teoh, A. P. (2019). The role of social media on recycling behaviour. Sustainable Production and Consumption, 20, 365–374.

Vaingankar, J. A., Van Dam, R. M., Samari, E., Chang, S., Seow, E., Chua, Y. C., … Subramaniam, M. (2022). Social media–driven routes to positive mental health among youth: Qualitative enquiry and concept mapping study. JMIR Pediatrics and Parenting, 5(1), e32758.

Wang, C., Li, Z., & Huang, X. (2018). Geospatial assessment of wetness dynamics in the October 2015 SC flood with remote sensing and social media. Southeastern Geographer, 58(2), 164–180.

Wang, S., Huang, X., Hu, T., She, B., Zhang, M., Wang, R., … Bao, S. (2023a). A global portrait of expressed mental health signals towards COVID-19 in social media space. International Journal of Applied Earth Observation and Geoinformation, 116, 103160.

Wang, S., Ning, H., Huang, X., Xiao, Y., Zhang, M., Yang, E. F., … Zeng, Y. (2023b). Public surveillance of social media for suicide using advanced deep learning models in Japan: Time series study from 2012 to 2022. Journal of Medical Internet Research, 25, e47225.

Wang, S., Zhang, M., Huang, X., Hu, T., Li, Z., Sun, Q. C., & Liu, Y. (2022a). Urban-regional disparities in mental health signals in Australia during the COVID-19 pandemic: A study via Twitter data and machine learning models. Cambridge Journal of Regions, Economy and Society, 15(3), 663–682.

Wang, Z., & Huang, W. (2020). A social media based approach for route planning during urban events. IEEE Access, 8, 207589–207598.

Wang, Z., Lam, N. S., Sun, M., Huang, X., Shang, J., Zou, L., … Mihunov, V. V. (2022b). A machine learning approach for detecting rescue requests from social media. ISPRS International Journal of Geo-Information, 11(11), 570.

Wang, Z., Miao, Y., Xu, M., Zhu, Z., Qureshi, S., & Chang, Q. (2021). Revealing the differences of urban parks' services to human wellbeing based upon social media data. Urban Forestry & Urban Greening, 63, 127233.

Williams, H. T., McMurray, J. R., Kurz, T., & Lambert, F. H. (2015). Network analysis reveals open forums and echo chambers in social media discussions of climate change. Global Environmental Change, 32, 126–138.

Yue, L., Chen, W., Li, X., Zuo, W., & Yin, M. (2019). A survey of sentiment analysis in social media. Knowledge and Information Systems, 60, 617–663.

7

Geospatial Optimization Coupling Spatiotemporal Big Data and Deep Reinforcement Learning

Shaohua Wang

7.1 Introduction

Geographic Information Systems (GIS) are technical systems, supported by computer hardware and software, that collect, store, manage, compute, analyze, display, and describe geographically distributed data within all or part of the Earth's surface (including the atmosphere), and space [1]. GIS combines geography and information technology sciences. Early GIS centered on cartography, utilizing computers for map drawing. It evolved to be data- and structure-centered, involving the modeling of geographic information data and the management of spatial databases. Modern GIS not only processes and analyzes geospatial data but also integrates these data with other relevant information to provide comprehensive solutions across various fields. With the development of technologies such as big data, artificial intelligence, and cloud computing, GIS applications are progressively becoming intelligent, automated, and cloud-based, offering more efficient and precise solutions to complex spatial problems. This evolution has significantly impacted social and economic activities [2, 3], establishing GIS as a crucial tool in promoting intelligent and digital transformation.

As GIS technology continues to advance, its application scope is consistently expanding, permeating various fields and playing a pivotal role in solving complex spatial optimization problems [4]. Particularly in resource allocation, optimizing the use and allocation of spatial resources is the primary focus of geospatial optimization research. Geospatial optimization refers to the utilization of GIS technology to optimize the allocation, utilization, and management of spatial resources, aiming to maximize resource utilization and enhance socioeconomic benefits. It involves analyzing spatial data and associated attribute data, constructing mathematical models, and employing optimization algorithms for problem-solving, ultimately assisting decision-makers in making optimal choices within complex geospatial environments.

DOI: 10.1201/9781003484219-7

Geospatial optimization is a hot interdisciplinary research area, intersecting geographic information science, remote sensing science, operations research, informatics, and humanities and social sciences. Research on geospatial optimization theories, methods, and technologies holds significant scientific importance and practical value. The integrated application of combinatorial optimization and GIS represents a classic geospatial optimization problem [5].

Spatial optimization problems are widely present across human, resource, and ecological spaces. Geospatial optimization enables the reconfiguration of resources, paths, locations, and facilities to maximize resource utilization. For instance, the siting of warehouses and wholesale stations typically aims to minimize costs [6]; the location of public service facilities such as parks, libraries, and stadiums needs to consider factors like proximity and equity [7]; whereas the placement of billboards, restaurants, and retail stores focuses on issues like foot traffic [8]. Specific practical problems can be addressed by establishing different models to achieve optimization.

The complexity of geospatial optimization problems lies in their involvement of multiple spatial dimensions and intricate interrelationships [9] (Figure 7.1). These problems can be described as optimization tasks requiring the establishment of mathematical models for solution. Such models usually comprise three components: objectives, decisions, and constraints. The objective function reflects the goals to be achieved, such as minimizing costs or maximizing benefits, and may involve one or multiple objective functions. Decision variables are the adjustable parameters in mathematical programming, corresponding to practical decisions to be made—for example, determining the placement of transmission corridors, the number

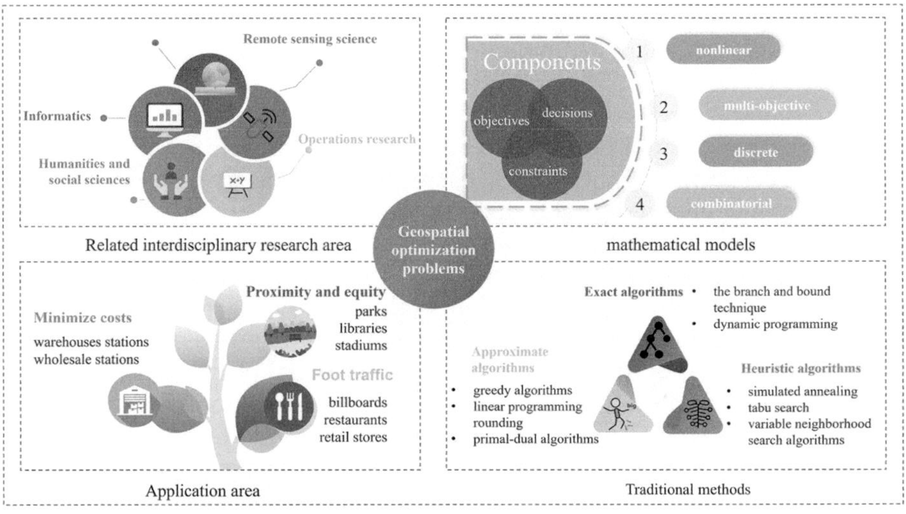

FIGURE 7.1
Multidimensional analysis diagram of geospatial optimization problems.

of sensors required to monitor an area, or when to open new supply distribution centers. Constraints are necessary conditions that must be satisfied in the problem under study. In mathematical programming, they restrict the values of variables, and in geospatial optimization problems, they represent geographical information and spatial limitations—such as excluding certain areas from infrastructure development or allocating more resources to specific regions. Tong and Murray summarized eight characteristics in geospatial optimization, including distance, adjacency, connectivity, containment, intersection, shape, area, and pattern. Incorporating geospatial optimization features can effectively reduce the solution space and expedite the problem-solving process [10].

Geospatial optimization problems typically exhibit high complexity, with mathematical models often demonstrating nonlinear, multi-objective, discrete, or combinatorial characteristics. This complexity renders these problems as NP-hard, with solution processes that grow exponentially with increasing problem size. Consequently, solving such problems often necessitates specific algorithms and strategies. Traditional methods for solving geospatial optimization problems mainly include exact algorithms, approximate algorithms, and heuristic algorithms [11]. Exact algorithms aim to find the optimal solution through exhaustive search, with representative methods such as the branch and bound technique [12] and dynamic programming. Approximate algorithms seek to find suboptimal solutions [13], including greedy algorithms, linear programming rounding, and primal-dual algorithms. Heuristic algorithms, constructed based on intuition or experience, can identify relatively good solutions within a short time frame. Common examples include simulated annealing, tabu search, and variable neighborhood search algorithms [14].

The inherent complexity of geographical problems presents significant challenges to solving spatial optimization issues. Developing exact and heuristic solution methods for spatial optimization problems is likely to remain an important and active research area. The subsequent two sections of this chapter will explore data- and model-driven geospatial optimization problems and geospatial optimization problems based on deep reinforcement learning (DRL).

7.2 Geospatial Optimization Research Driven by Data and Model Coupling

In geospatial optimization problems, the effective integration of data and models is the key to achieving efficient optimization. Geospatial optimization relies not only on a profound understanding of complex spatial relationships but also on accurate and rich data support. Data not only provide the basic inputs and constraints for optimization models but also, by describing the diversity and complexity of geographical phenomena, help models align more closely with real-world needs. Meanwhile, optimization models offer scientific frameworks and algorithms for data utilization, ensuring that optimal or near-optimal solutions can be obtained under multi-objective and multi-constraint scenarios.

The close coupling between data and models makes the geospatial optimization process more precise and efficient. The quality, accuracy, and timeliness of data directly affect the reliability of optimization results, while the scientific rigor and adaptability of models determine the effectiveness of the optimization process. By organically combining data and models, researchers can make wiser and more efficient decisions in complex geospatial contexts. This coupling not only enhances the predictive power of models but also allows for dynamic adjustment of optimization strategies through real-time data updates and feedback, thereby responding to the constantly changing geographical environment.

Next, we will delve into geospatial optimization research driven by data and model coupling. First, we introduce classic geospatial optimization location models, then explore the application of these models in specific cases (Figure 7.2). To facilitate readers' understanding of the mathematical models, we will use unified symbol definitions. In geospatial location optimization problems, there is usually a set of demand points I, representing all demand points i. Each demand point has a corresponding weight or demand ω_i, which influences facility siting. The set of candidate facility sites is denoted as J, representing all possible facility locations j. The number of facilities to be located is determined as p, and each facility has a corresponding service coverage

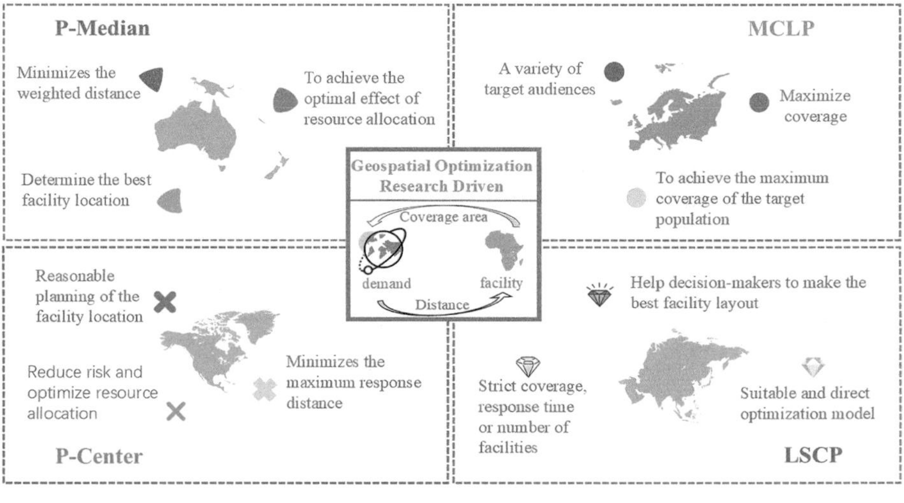

FIGURE 7.2
Facility layout model comparison driven by geospatial optimization research.

range S. When $d_{ij} \leq S$, facility j can cover demand point i. The distance (cost) from demand point i to facility j is denoted as d_{ij}. To distinguish whether a facility is selected, we set a decision variable x_j; if a facility is located at candidate site j, then $x_j = 1$; otherwise, $x_j = 0$. Whether a demand point is served by a facility is represented by the decision variable y_{ij}; if demand point i is served by facility j, then $y_{ij} = 1$; otherwise, $y_{ij} = 0$. Additionally, in the p-Center problem, we need to define Z to represent the maximum distance between facilities and the demand points they cover; z_i is a coverage variable, where $z_i = 1$ if demand point i is covered; otherwise, $z_i = 0$.

7.2.1 Balancing Optimal Facility Location and Coverage

The balance between optimal facility location and coverage (the p-Median problem) aims to optimize facility placement by selecting p facility sites to minimize the total distance from demand points to their nearest facilities, thereby maximizing cost-effectiveness. This problem is widely applied in public facility siting, network design, logistics management, and other fields. To solve the p-Median problem, a common method is to transform it into a Mixed Integer Linear Programming (MILP) problem. By introducing decision variables and constraints, the objective function and constraints of the p-Median problem can be formulated in the following MILP form:

$$\sum_{i \in I} \sum_{j \in J} \omega_i \cdot d_{ij} \cdot y_{ij} \tag{7.1}$$

$$\sum_{j \in J} x_j = p \tag{7.2}$$

$$\sum_{j \in J} y_{ij} = 1, \forall i \in I \tag{7.3}$$

$$y_{ij} \leq x_j, \forall i \in I, j \in J \tag{7.4}$$

$$x_j \in \{0,1\}, y_{ij} \in \{0,1\}, \forall i \in I, j \in J \tag{7.5}$$

Where Eq. (7.1) is the objective function, aiming to minimize the total weighted distance from all demand points iii to the selected facility sites j; constraint Eq. (7.2) stipulates that a total of p facility sites must be selected; constraint Eq. (7.3) ensures that each demand point i must be served by a facility site p; in constraint Eq. (7.4), only when facility site j is selected as a facility can demand point iii be served by it. The variables x_j and y_{ij} are binary variables, representing whether to locate a facility at site j and whether demand point i is served by facility site j, respectively.

The p-Median problem is widely applied in multiple fields, especially advantageous in solving location problems. By minimizing the weighted distances between facilities and demand points, it helps decision-makers determine the optimal facility locations to achieve the best effect of resource allocation. Its application scenarios include, but are not limited to, transportation, logistics, and public services.

In the siting of social service facilities, government agencies utilize the p-Median problem to determine the locations of community service institutions such as libraries, medical clinics, and police stations, to provide the most convenient services to residents. Taking medical facility siting as an example, government agencies or private medical institutions can use the p-Median problem to determine the locations of new medical facilities, allowing more residents to access medical services more conveniently [15]. Additionally, in the field of education, schools can also use the p-Median problem to determine the locations of new campuses or teaching sites, minimizing the distance students need to travel to the nearest teaching point [16].

Logistics companies can use the p-Median problem to determine the locations of new distribution centers, minimizing the total distance from customers to their nearest center, thereby improving delivery efficiency [17]. In the siting of public service facilities, the p-Median problem is adopted to determine the locations of public transportation stations, allowing as many residents as possible to access public transportation services more conveniently [18]. Moreover, government agencies use this to determine the locations of new power generation stations, aiming to minimize power transmission costs while considering environmental factors and social impacts [19].

In daily life, there is a type of siting known as undesirable facility location, referring to facilities that may pollute the environment, harm or potentially harm people's physical and mental health, such as nuclear power plants, waste treatment plants, chemical factories, and mobile base station towers. These undesirable facilities are often necessary public facilities, but people usually prefer to stay away from them. Taking waste treatment facility siting as an example, government agencies can use the p-Median problem to determine the locations of new waste treatment facilities, minimizing the total distance from waste sources to the nearest facilities, thereby improving waste processing efficiency. In this process, by considering the distance between each resident and their nearest acceptable facility and the penalty degree for residents lacking acceptable facilities, the undesirable facility location problem can be effectively solved, finding the optimal combination of facility locations to meet requirements and minimize the total distance [20].

The charging station layout problem is another application area of the p-Median problem. By minimizing the total distance from all electric vehicles to their nearest acceptable charging stations, a set of charging station locations can be determined. Considering the distance between each electric vehicle and its nearest acceptable charging station, and treating the penalty degree for electric vehicles lacking acceptable charging stations as infinitely large, by solving the *p*-Median problem, the optimal combination of charging station locations can be found to meet the requirements and minimize the total distance [21].

7.2.2 Maximizing Service Coverage and Emergency Response Capability

The optimization strategy for maximizing service coverage and emergency response capability (the *p*-Center problem) studies how to locate *p* facilities within a network. This problem was first proposed by Hakimi [13] in 1964, who introduced the absolute center problem for location optimization to determine the placement of police stations or hospitals. The sites determined by this method can minimize the maximum distance between a set of communities connected by a road system; therefore, it is also known as the minimax problem. The objective function and constraints of the *p*-Center problem can be formulated as the following MILP:

$$Z \tag{7.6}$$

$$\sum_{j \in J} x_j = p \tag{7.7}$$

$$d_{ij} \cdot y_{ij} \le Z, \forall i \in I, j \in J \tag{7.8}$$

$$\sum_{j \in J} y_{ij} = 1, \forall i \in I \tag{7.9}$$

$$y_{ij} \le x_j, \forall i \in I, j \in J \tag{7.10}$$

$$x_j \in \{0,1\}, y_{ij} \in \{0,1\}, Z \ge 0, \forall i \in I, j \in J \tag{7.11}$$

The optimization objective function Eq. (7.6) of the *p*-Center problem is to minimize the maximum service distance Z, which is the greatest distance among the demand points covered by the facilities; constraint Eq. (7.7) indicates that the number of facility sites is fixed at p; constraint Eq. (7.8) states that for each demand point i, the distance d_{ij} to the facility site j serving it must not exceed the maximum distance Z. By minimizing Z, we ensure that the maximum service distance is as small as possible; each demand point i must be served by a facility site j (constraint Eq. (7.8)), and only when facility site j is selected as a service point can demand point i be served by it (constraint Eq. (7.9)); x_j and y_{ij} are binary variables, and Z is a non-negative real number representing the maximum service distance.

The *p*-Center problem has a wide range of applications, mainly focusing on scenarios that require minimizing the maximum response distance or time. By reasonably planning facility locations, the *p*-Center problem can help improve service efficiency, reduce risks, and optimize resource allocation, and is widely used in emergency response, logistics distribution, public facilities, communication networks, and other fields.

In emergency response, the *p*-Center problem can be applied to determine the optimal locations of emergency response facilities such as fire stations and police stations, to ensure that the response time to any emergency incident is minimized [22]. By minimizing the maximum distance between demand points and service facilities, the efficiency of emergency services can be improved, reducing the losses caused by disasters and accidents [23]. In medical emergencies, the *p*-Center problem can help plan the layout of ambulance stations, ensuring that emergency vehicles can reach accident sites in the shortest possible time, thereby increasing the success rate of rescues.

Apart from application research, solving such problems is also very important. Minieka proposed the first algorithm for solving the discrete *p*-Center problem based on a set covering problem with a finite set of points. Garfinkel et al. [24] improved the set covering method based on Minieka's work by first finding a heuristic solution, then reducing the search space of radius values and eliminating some intersections, and reducing the size of the set covering matrix by using standard matrix reduction and heuristic techniques. Elloumi et al. proposed a new IP formulation to solve the *p*-Center problem, utilizing the fact that the optimal value of the *p*-Center problem is constrained by a finite set of distance values [25]. Daskin provided a mixed integer programming (MIP) formulation for the discrete *p*-Center problem and proposed a set covering-based algorithm to solve it, where the radius values are chosen from a real interval between predetermined lower and upper bounds [26]. Mladenović et al. were the first to propose metaheuristic methods for approximately solving the *p*-Center problem, including Multi-start Local Search Algorithm (M-I), Tabu Search Algorithm (TS) with Chain Exchanges, and Variable Neighborhood Search Algorithm (VNS). The proposal of these methods provided an effective approach to solving the *p*-Center problem [27]. Pullan proposed a memetic algorithm for the vertex-restricted *p*-Center problem, combining population-based metaheuristic algorithms with local search algorithms to find optimized solutions to the *p*-Center problem; by utilizing information exchange and genetic operations among individuals in the population, exploration of the solution space, and optimization of the search process were achieved [28].

These algorithms and methods provide diversified strategies for solving the *p*-Center problem. From local centers, set covering, integer programming to metaheuristic methods and memetic algorithms, they continuously promote the research and application of the *p*-Center problem. The proposal and improvement of these methods provide powerful tools and ideas for solving practical problems such as facility location and resource allocation.

7.2.3 Maximizing Demand Coverage Optimization Strategy

The Maximizing Coverage Location Problem (MCLP) seeks to maximize the number of demand points within the service coverage area by selecting a limited number of facility locations. The core objective of MCLP is to utilize limited resources so that as many demand points as possible can be served by at least one facility. This problem is widely applied in public service siting, emergency facility layout, and commercial site selection, enhancing service coverage and overall effectiveness by optimizing facility locations. The mixed-integer linear programming formulation of MCLP is as follows:

$$Max \sum_{i \in I} \omega_i \cdot z_i \tag{7.12}$$

$$\sum_{j \in J} x_j = p \tag{7.13}$$

$$\sum_{j \in J,\, d_{ij} \le S} y_{ij} \ge z_i, \forall i \in I \tag{7.14}$$

$$y_{ij} \le x_j, \forall i \in I, j \in J \tag{7.15}$$

$$x_{ij} \in \{0,1\}, y_{ij} \in \{0,1\}, z_{ij} \in \{0,1\}, \forall i \in I, j \in J \tag{7.16}$$

The optimization objective Eq. (7.12) of MCLP is to maximize the amount of demand covered; similar to the previous *p*-Median problem and *p*-Center problem; constraint Eq. (7.13) determines the number of facility sites as *p*; in constraint Eq. (7.14), if demand point *i* is covered by at least one facility site (i.e., $d_{ij} \le S$), then $z_i = 1$, indicating that the demand point is covered; constraint Eq. (7.15) states that only when facility site *j* is selected as a service point can demand point *i* be served by that facility site. x_j, y_{ij}, and z_i are binary variables. z_i indicates whether demand point *i* is covered.

MCLP is widely applied in commercial location selection, such as billboard placement and digital signage siting. By using MCLP, it is possible to achieve maximum coverage of target audiences, maximum traffic flow coverage, and a balance among multiple target demographics. Furthermore, MCLP can handle complex real-world constraints in billboard layouts, such as geographical restrictions, policies and regulations, and interference from competitors. By reasonably applying the MCLP model, the allocation of advertising resources can be optimized as much as possible to ensure the maximization of advertising effectiveness [29].

In the siting of medical facilities, MCLP can also be used for modeling because medical resources are often limited. How to optimize the layout under limited resources to maximize the coverage of surrounding areas' demand for medical resources, ensuring the effectiveness and response speed of emergency services, is something MCLP can solve. Through MCLP modeling, the allocation of medical resources can achieve optimal configuration, enabling more residents to receive responses in the shortest possible time when emergency needs arise [30].

Many scholars have also conducted considerable research on algorithms for solving MCLP, such as genetic algorithms that simulate the concept of biological evolution for optimization. Genetic algorithms are suitable for large-scale problems but may have long running times. To address this, Han and Li [30] proposed an improved adaptive genetic algorithm concerning computation time and resources. In solving the MCLP problem, using genetic algorithms to cover the maximum population within a given distance of public facilities [31], using genetic algorithms to maximize customer service with limited resources [32], and using genetic algorithms to construct location-allocation models in medical facility planning [33] are all very typical optimization algorithms.

7.2.4 Method for Ensuring Minimum Service Coverage Demand

The Method for Ensuring Minimum Service Coverage Demand (LSCP) is a classic location optimization problem. It aims to ensure that all demand points can be covered by selecting the minimal number of facility locations. LSCP is widely applied in scenarios such as public service siting, emergency response facility layout, and communication network design, helping to maximize service coverage under limited resources. The core objective of this problem is to minimize the number of facilities while satisfying the coverage requirements of all demand points.

$$\sum_{j \in J} x_j \tag{7.17}$$

$$\sum_{j \in J,\, d_{ij} \le S} x_j \ge 1, \forall i \in I \tag{7.18}$$

$$x_j \in \{0,1\}, \forall j \in J \tag{7.19}$$

LSCP is a classic location optimization problem, aiming to ensure that all demand points can be covered by selecting the minimal number of facility locations. LSCP is widely applied in public service siting, emergency response facility layout, and communication network design, helping to maximize service coverage under limited resources. The core objective of this problem is to minimize the number of facilities while satisfying the coverage requirements of all demand points.

The optimization objective of LSCP is to minimize the number of facility sites, covering all demand points with as few facility sites as possible Eq. (7.17);

constraint condition Eq. (7.18) stipulates that each demand point i must be covered by at least one facility site (i.e., facility sites where $d_{ij} \leq S$); x_j is a binary variable indicating whether facility site j is selected.

The LSCP model provides a basic framework and method for solving location set covering problems and has led to further exploration of optimization algorithms and heuristic strategies. In addition, the LSCP model has found widespread application cases in many fields, often in scenarios with strict requirements on coverage area, response time, or number of facilities. In these scenarios, LSCP is a very suitable and straightforward optimization model that can help decision-makers make the best facility layout choices.

The siting of electric vehicle charging stations is an application scenario with a limited number of facilities; reasonable charging station placement will increase the driving range of electric vehicles. Through the LSCP model, planners can achieve the optimal layout of electric vehicle charging stations under limited resource conditions, ensuring that all car owners can receive good charging services. This siting method can help improve the coverage rate of the charging network, reduce users' charging anxiety, and promote the popularization of electric vehicles [34].

The siting of post-disaster emergency centers can be categorized as an LSCP with requirements on both coverage area and response time, and generally, the number of emergency centers is limited. Under a limited number of facilities, optimizing the layout to maximize service coverage ensures that all affected people can receive timely help. In flood disasters, large areas are usually submerged, and some roads are cut off. The LSCP model can layout emergency centers on high ground or safe areas not submerged by floods, ensuring that these emergency centers can cover surrounding residential areas affected by the flood [35].

Daskin proposed the MECLP model to solve the optimal response problem in overloaded emergency service systems [36]; Wu et al. achieved optimal coverage of water level sensors in mines using the LSCP model and greedy algorithms [37]; and Zhong Qing and Dai Teqi conducted optimal coverage analysis of airport distribution scales through a comprehensive correction model combining LSCP and MCLP [38].

The solution methods for the LSCP problem are mainly divided into exact solutions and heuristic solutions. The most commonly used exact algorithm is branch and bound, which searches for the solution on the problem's solution space tree. Etcheberry used a strategy based on branch and bound to solve the LSCP problem, but its branching strategy is aimed at constraints rather than variables, calculating the lower bounds of the tree using the Lagrangian relaxation method [39]. When the number of demand points is large, the speed of using exact methods will be very slow. If some application scenarios have high real-time requirements for solving the problem, then heuristic algorithms need to be considered.

With the definition of new models, improvements to existing models, and the emergence of more challenging problem instances, solving problems will continue to pose challenges to spatial optimization applications. It's not just a matter of using exact methods or heuristic methods; this indicates that the development of solutions and the continuous pursuit of better methods will always be important. However, spatial optimization researchers also face other challenges, such as abstraction and representation issues, the identification of multiple problem objectives and the need to consider alternative optima, and further utilization of topological relationships obtained through GIS.

Similar to other methods in geography, spatial optimization problems are often constrained by the Modifiable Areal Unit Problem (MAUP) [40] in terms of analysis scale and unit definition. In the application of the *p*-Median problem, Fotheringham et al. emphasized that solutions are highly dependent on how geographical units are defined and recommended caution when using aggregated spatial units [41]. Although Murray and Gottsegen's research suggests this is somewhat exaggerated, the problem still exists [42]. Horner and Murray used transportation issues to illustrate the complexity of scale and unit definition when evaluating urban excessive commuting [43]. Murray and O'Kelly provided examples of abstraction and MAUP problems in coverage optimization [44], and Cromley and Mrozinski also provided supporting evidence in this regard [45].

The difficulty of geographical unit modeling and solving related problem instances poses challenges to spatial optimization. To reduce or mitigate errors and uncertainties in the geospatial abstraction process, extensive research is needed. One potential research direction to solve this problem is to develop new abstraction schemes that more realistically represent geographical units. Recent studies by [46] and [10] have shown prospects in this regard. However, new spatial abstraction strategies will undoubtedly require the development of new models and more efficient methods to solve these problems (Table 7.1).

TABLE 7.1

Geospatial Optimization Research Driven

Space Optimization Model	Advantage
p-Median	Minimizes the weighted distance; To achieve the optimal effect of resource allocation; Determine the best facility location
p-Center	Reasonable planning of the facility location; Reduce risk and optimize resource allocation; Minimizes the maximum response distance
MCLP	To achieve the maximum coverage of the target population; A variety of target audiences; Maximize coverage
LSCP	Help decision-makers to make the best facility layout; Strict coverage, response time, or number of facilities; Suitable and direct optimization model

Due to the inherent spatial background of geospatial optimization problems, the importance of GIS is increasingly recognized, especially in spatial data management and analysis. Traditional optimization methods (such as exact methods and heuristic methods) often face bottlenecks when dealing with large-scale, complex geospatial problems. This provides new opportunities for the application of emerging technologies such as deep learning in geospatial optimization. Utilizing advanced models like deep neural networks (DNNs) and graph neural networks, deep learning can automatically extract and model complex relationships in spatial data, overcoming the limitations of traditional methods in feature engineering and nonlinear problems. In addition, deep learning can more effectively handle multi-objective optimization problems in dynamic environments, thereby promoting geospatial optimization towards more efficient and intelligent development.

7.3 Research on Geospatial Optimization Based on Deep Reinforcement Learning

7.3.1 Applicability of Deep Reinforcement Learning in Geospatial Optimization

With the advent of the big data era and the continuous improvement of computational power, artificial intelligence has achieved unprecedented development [47]. DRL is an artificial intelligence technique that combines deep learning and reinforcement learning. The fundamental principle of reinforcement learning is to learn a policy through continuous interaction with the environment, aiming to maximize accumulated rewards by selecting a sequence of actions. In this process, an agent observes the current state, uses the policy to decide the next action, receives reward feedback from the environment, and adjusts the policy to achieve higher long-term returns. The introduction of deep learning enables reinforcement learning to handle more complex input data, such as images and high-dimensional feature vectors, by using DNNs for automatic feature extraction and nonlinear mapping.

In geospatial optimization problems, DRL can help build more intelligent decision models by automatically learning and optimizing complex relationships and latent patterns in geospatial data. Traditional reinforcement learning methods often exhibit limitations when facing high-dimensional and continuous state spaces, whereas deep learning can leverage its strong representation capabilities to process complex geospatial data (e.g., satellite images, spatial networks, and spatiotemporal sequence data) and generate effective decision strategies.

In geospatial optimization problems, the high dimensionality and complexity of data are often the main challenges in the solution process. Traditional geospatial optimization methods rely on predefined models and parameters, the design of which often requires extensive domain knowledge and may need to be readjusted or redesigned when facing different application scenarios. DRL has adaptive learning capabilities; it updates its policies through continuous interaction with the environment without the need for manually designed complex model structures, thereby better adapting to different geospatial optimization problems and application scenarios. Especially when dealing with multi-objective optimization and uncertain environments, DRL can gradually learn optimal or suboptimal solutions through policy iteration and value function approximation. Geospatial data usually exhibit spatiotemporal correlation and diversity, such as meteorological data, traffic flow data, and land use data. DRL utilizes DNN to perform feature extraction and pattern recognition on these complex input data, automatically discovering useful information hidden within the data, and enhances geospatial optimization effects through this information. Many geospatial optimization problems (e.g., traffic management, logistics distribution, disaster response) are dynamic and changing, which traditional static optimization models struggle to handle effectively. DRL, through continuous learning and adjustment, can adapt in real-time to environmental changes, providing more robust optimization decisions.

In the field of geospatial optimization, traditional optimization methods like MILP models have a rigorous theoretical foundation and good solution accuracy when dealing with small-scale or specifically structured problems. However, these methods often face issues such as long computation times and difficulty in solving when confronted with large-scale, nonlinear, or uncertain complex problems. On the other hand, heuristic algorithms (e.g., genetic algorithms, simulated annealing algorithms) can alleviate computational complexity to some extent, but their solution quality usually depends on precise parameter tuning and may struggle to achieve ideal results in multi-objective and dynamic optimization problems.

Compared to these traditional methods, the greatest advantage of DRL lies in its model-free nature. It does not rely on precise mathematical descriptions of the problem but instead automatically approximates the optimal policy through continuous interaction and learning with the environment. This characteristic allows DRL to perform excellently in complex and highly uncertain problems. Moreover, DRL can handle high-dimensional input data, especially when dealing with geospatial optimization problems involving multiple data sources and features, demonstrating strong flexibility and scalability. In terms of computational efficiency, although DRL may require significant computational resources during initial training, once training is complete, the model's inference speed is relatively fast, making it suitable for geospatial applications that require real-time decision-making. In contrast, traditional optimization methods often need to compute from scratch each time the problem is solved, which is less efficient in dynamically changing environments.

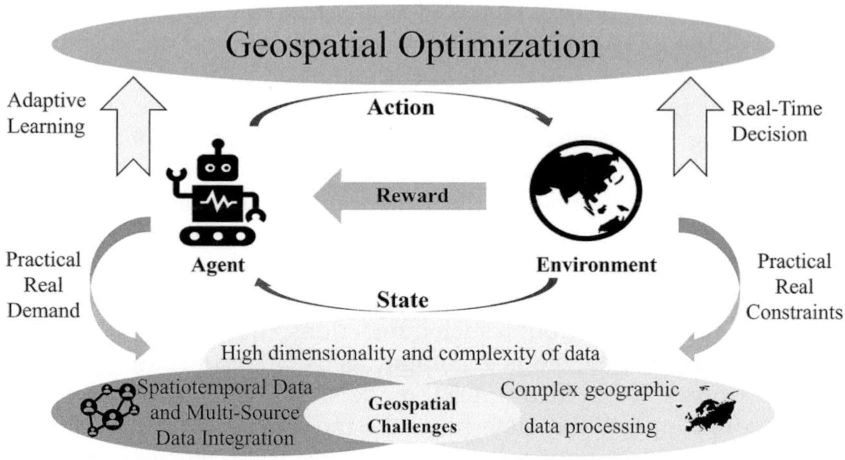

FIGURE 7.3
A combined model of deep reinforcement learning and geospatial optimization.

DRL offers a completely new solution approach for geospatial optimization problems, showcasing tremendous application potential through advantages such as adaptive learning, efficient feature extraction, and model-free problem-solving (Figure 7.3). Although DRL still faces some challenges in specific applications, such as large training data requirements and insufficient model interpretability, its application prospects in the field of geospatial optimization will become increasingly broad with technological advancements and the emergence of more application cases.

7.3.2 Construction of Geospatial Optimization Models Driven by Deep Reinforcement Learning

In DRL, the definition of the agent and the environment is the foundation for constructing effective geospatial optimization models. The agent is the subject that executes actions and learns policies, while the environment is the external system in which the agent operates, providing corresponding rewards and updating states based on the agent's actions.

7.3.2.1 Construction of Deep Reinforcement Models

In geospatial optimization problems, the agent usually represents the decision-maker or optimization algorithm, while the environment can be various dynamic systems in the geospatial domain, such as transportation networks, logistics distribution networks, and urban planning areas. In a *p*-Median problem, the agent can be defined as the decision-maker responsible for selecting facility locations, whose task is to choose several points from a set of candidate locations to minimize the total weighted distance from all demand points to the nearest facility. The environment is the geospatial area composed of the distribution of demand points and candidate facility locations, containing information such as the geographical positions of demand points, their weights, and the selectable positions of candidate facilities. The key to defining the agent and environment lies in clearly delineating the interaction mechanism between them. Each action of the agent (e.g., selecting a facility location) will affect the state of the environment (e.g., changing the distribution of distances from demand points to the nearest facility), and the environment will provide the agent with a reward based on the new state (e.g., the negative value of the total distance, indicating the improvement of the objective function). Through such interactions, the agent gradually learns how to make optimal decisions in complex geospatial contexts.

In the process of constructing DRL models, the state, action, and reward are three core elements that determine the model's performance and learning effect. In geospatial optimization problems, these elements need to be carefully designed.

The state space represents the environmental information observed by the agent at any given time. In geospatial optimization problems, the state usually includes the current distribution of facilities, the distribution and demand of demand points, the availability of candidate locations, and possible temporal factors (such as time periods and seasonal variations). In the p-Center problem, the state can include the set of currently selected facility locations and the distance from each demand point to the nearest facility. This information helps the agent assess the effectiveness of current decisions and the next action. The design of the state space should consider the completeness and compactness of the information, covering key factors that affect decision-making while avoiding redundant information to reduce computational complexity. Additionally, given the high dimensionality of geospatial data, DNNs (such as Convolutional Neural Networks or Recurrent Neural Networks) can be used to extract spatial and temporal features, simplifying state representation.

The action space defines the operations that the agent can take in each state. In geospatial optimization problems, actions can be selecting or removing a facility location or adjusting resource allocation in a certain area. In the p-Median problem, actions might involve selecting a location from the candidate set as a new facility point or removing a location from the currently selected facilities. The size of the action space directly affects the complexity and solution efficiency of the DRL model. Therefore, methods such as limiting the number of candidate locations, introducing hierarchical decision structures, or using continuous action spaces can be employed to reduce the scale of the action space, thereby accelerating the model's convergence process.

The reward mechanism is the core of reinforcement learning; it directly drives the agent's learning process. In geospatial optimization problems, rewards are usually directly related to the optimization objectives. For the *p*-Median problem, the reward can be set as the amount of reduction in the total weighted distance after each action; for the *p*-Center problem, the reward can be linked to the objective function of minimizing the maximum distance. The core idea in reward design is to encourage the agent to continuously improve toward the optimization goal and provide clear guiding signals after each operation. To prevent the model from falling into local optima, the reward mechanism should also include some exploratory rewards to encourage the agent to try different strategies. For example, a small positive reward can be given for each new facility location selected to encourage the agent to explore more possibilities. Additionally, in some dynamic optimization problems, the reward mechanism should consider temporal factors, giving higher rewards to strategies that achieve the goal more quickly.

7.3.2.2 Training and Optimization of Deep Reinforcement Models

Once the agent, environment, state, action, and reward mechanism are clearly defined, the next step is to train the DRL model to gradually approach the optimal policy.

Training of DRL models usually requires a large amount of data. In geospatial optimization, this data includes historical demand data, geographic information data (spatial information data), traffic flow data, etc. Existing spatial databases can be utilized, or training data can be obtained through simulation generation (e.g., based on multi-agent simulation systems). For dynamic optimization problems, it is also necessary to consider the acquisition and processing of time-series data so that the model can learn patterns that change over time.

During training, the hyperparameter settings of the model (learning rate, discount factor, exploration strategy, etc.) have a significant impact on the final convergence speed and effect. Typically, tuning is required through experiments. Since models based on deep learning have a large number of training parameters and complex training processes, large-scale parallel computing platforms can be used to accelerate training. Moreover, to prevent model overfitting, techniques such as experience replay and target networks can be employed to enhance the model's generalization ability.

Different reinforcement learning algorithms may exhibit different advantages and disadvantages in geospatial optimization problems. For example, DQN is suitable for problems with discrete action spaces, while algorithms like Proximal Policy Optimization (PPO) and A3C are more effective in continuous action spaces. The choice of an appropriate algorithm depends on the specific characteristics of the problem, such as the nature of the action space, the degree of environmental dynamics, and the availability of computational resources.

During training, the model continuously interacts with the environment, gradually updating the policy to approximate the optimal solution. When the model's policy stabilizes and performs well on the validation dataset, the training can be considered complete, and the model can be used in actual geospatial optimization tasks. At this point, the model's output will be a policy that can be directly applied to practical decision-making processes, capable of providing optimal geospatial layouts or resource allocation schemes in real time.

7.3.3 Applications of Deep Reinforcement Learning in Geospatial Optimization

In geospatial optimization problems, the locations of demand points are usually fixed, while the layout of facility points can be flexibly adjusted using various schemes, reflecting the dynamic spatial allocation of facility services. This type of optimization layout problem can generally be divided into facility location and service coverage, with the optimization goal being to adjust the layout of facility points in time and space to ensure the effective utilization and balanced distribution of facility services among various demand points. In this process, the allocation of facility services is based on constraints and follows certain rules to cover as much demand as possible with limited facility services. In real life, such optimization problems are widely applied in fields such as power grid energy distribution [48, 49], medical service allocation [50], and rational layout of educational resources.

In the field of energy supply, the location optimization of urban electric vehicle charging stations is a hot research topic. Fast-charging stations for electric vehicles impose burdens on the power grid; therefore, optimizing the energy supply methods of charging stations is crucial while meeting the charging demands of electric vehicles. Heo and Chang used the Advantage Actor-Critic (A2C) DRL method to achieve optimal location optimization of electric vehicle fast-charging stations combined with solar energy supply in large and complex urban environments [51]. Padmanabhan et al. considered factors such as nearby traffic density, electric vehicle registrations, and distances to certain types of public buildings in charging pile siting, and combined DRL to establish a siting framework that can predict charging demand and current charging locations to find the optimal locations for new charging stations [52]. Zhao et al. formulated the fast-charging station deployment problem as a finite discrete Markov Decision Process within the reinforcement learning framework, proposed a Recurrent Neural Network with an attention mechanism to learn model parameters, and determined the optimal strategy for charging pile siting in an unsupervised manner [53]. Liu et al. incorporated an attention layer into the PPO algorithm, proposing a DRL model called PPO-Attention for optimizing charging station siting in road networks, considering factors such as the fairness and coverage of charging station distribution in their model [54].

In commercial site selection, the location of digital billboards directly affects advertising exposure, target audience coverage, and the actual effect of the advertisement. The formulation of siting strategies needs to consider multiple influencing factors such as population, traffic flow, competitive environment, and cost. To maximize revenue, Lou et al. used mobile crowd-sensing technology to collect vehicle data, extracted vehicle trajectories and preference information to identify potential customers, and proposed an advertising strategy based on multi-agent DRL [55]. Zhong et al. constructed a new MCBLP model based on MCLP for the billboard siting problem to maximize traffic flow coverage, developed a ReCovNet to solve MCBLP, integrated an attention mechanism into this network, and used real data from New York City for siting optimization [56]. Applications in commercial site selection also include optimizing station deployment for shared electric scooters using DRL [57].

To facilitate solving urban facility location problems, Liang et al. proposed a unified DRL framework, SpoNet, for solving p-Median, p-Center, and MCLP problems. SpoNet is more accurate than heuristic methods and much faster than solvers, with solution times improved by more than 20 times [58]. The layout of emergency medical facilities in cities is also crucial for urban development. Zhao and Wang used DRL to construct an emergency resource allocation system framework, optimized the system using the Deep Q Network path planning algorithm, and provided decision support for efficient scheduling of urban emergency resources [59]. Wang et al. proposed a MILP method for emergency logistics scheduling in flood emergency material distribution problems, used the DRL algorithm Deep Deterministic Policy Gradient (DDPG) to search for solutions, and greatly improved the solution efficiency [60].

The applications of DRL in geospatial optimization are continually expanding, from energy supply to commercial site selection and urban emergency management (Figure 7.4). Research in various fields has demonstrated that DRL can effectively address complex location and service coverage problems. With technological advancements and in-depth research, DRL will showcase its advantages in more practical geospatial optimization problems, providing more intelligent and efficient solutions for sustainable social development.

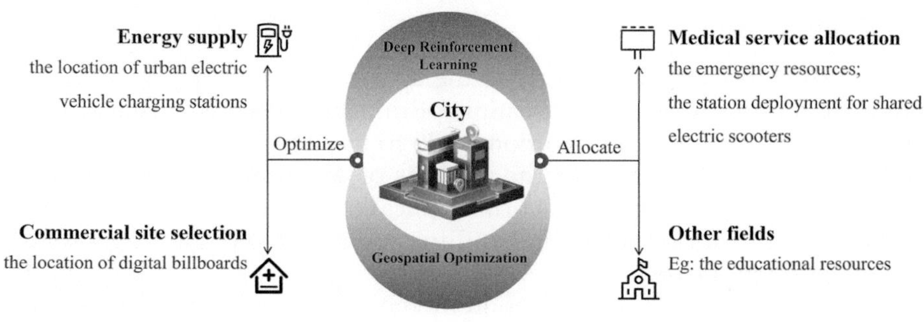

FIGURE 7.4
Urban geospatial optimization through deep reinforcement learning.

7.4 Discussion

This chapter thoroughly explores data- and model-driven geospatial optimization problems and geospatial optimization research based on DRL, revealing the potential and challenges of DRL in addressing complex and dynamic geospatial optimization issues. Compared with traditional optimization methods, DRL exhibits stronger adaptability and scalability, especially outstanding in handling large-scale, dynamic, and multi-objective optimization problems.

The application of DRL in geospatial optimization demonstrates significant advantages. DRL can handle high-dimensional data inputs and continuously learn and adapt in dynamic environments, ensuring the real-time effectiveness of optimization solutions. By combining multiple data sources (such as traffic flow, population distribution, resource demand, etc.), DRL can find better solutions in complex scenarios. Its adaptability enables it to rapidly adjust decisions in the face of uncertainty and dynamic changes, thereby enhancing the robustness of the optimization process.

Despite the many advantages of DRL in geospatial optimization, its application still faces some challenges. The model training process often requires a large amount of computational resources and time, especially under high-dimensional data and complex environments, where the convergence speed of the model is relatively slow, potentially leading to excessively high training costs. Due to the complexity and heterogeneity of geospatial data, the generalization ability of DRL models is limited; the models may not exhibit the same optimization effects in scenarios beyond the training data. In multi-objective optimization problems, how to reasonably design reward mechanisms to balance conflicts between different objectives is also a major challenge in DRL applications.

7.5 Conclusion

This chapter systematically reviews the application of coupling spatiotemporal big data with DRL in geospatial optimization research, discussing the evolutionary process from traditional GIS to modern DRL methods in geospatial optimization. Through the analysis of different optimization problems and the discussion of practical application cases, this chapter reveals the unique advantages of DRL in handling complex and dynamic geospatial optimization problems.

Traditional geospatial optimization methods, such as *p*-median, *p*-center, MCLP, and LSCP, although performing excellently in specific scenarios, often face limitations in computational resources and solution efficiency when dealing with large-scale, dynamically changing complex scenarios. Despite their solid foundations in mathematical modeling and optimization theory, these traditional methods seem inadequate when faced with increasingly complex real-world demands.

The introduction of DRL brings new solutions to geospatial optimization. By simulating the decision-making process in dynamic environments, DRL can continuously learn and adapt, handle uncertainty and multi-objective conflicts, and significantly improve the efficiency and effectiveness of optimization. The application examples in Chapter 3 show that DRL can provide more precise and faster optimization solutions than traditional methods in practical fields such as energy distribution, commercial site selection, and emergency management, especially demonstrating obvious advantages when dealing with complex multi-objective optimization problems.

By conducting an in-depth exploration of geospatial optimization problems, this chapter envisions its potential in solving complex optimization problems in the real world. With continuous technological development, DRL is expected to play a key role in more practical applications, opening new directions for geospatial optimization research, and providing more intelligent and efficient solutions for fields such as smart cities and resource management.

References (science advance)

1. Goodchild, M.F. Geographical Information Science. *International Journal of Geographical Information Systems*, 1992, *6*, 31–45, doi:10.1080/02693799208901893.
2. Zhong, E. GeoControl and Live Geography: Some Thoughts on the Direction of GIS. *Journal of Geo-Information Science*, 2013, *15*(6), 783–792.
3. Zhong, E. New Opportunities in the Development of Geographic Information Systems. *Frontiers of Science*, 2015, *9*(4), 19–28.
4. Murray, A.T. Contemporary Optimization Application Through Geographic Information Systems. *Omega*, 2021, *99*, 102176, doi:10.1016/j.omega.2019.102176.
5. Church, R.L.; Murray, A.T. *Business Site Selection, Location Analysis and GIS*; 1st ed.; Wiley, 2008.
6. Lee, B.K.; Kang, K.H.; Lee, Y.H. Decomposition Heuristic to Minimize Total Cost in a Multi-Level Supply Chain Network. *Computers & Industrial Engineering*, 2008, *54*, 945–959, doi:10.1016/j.cie.2007.11.005.
7. Peng, J.; Liu, Y.; Ruan, Z.; Yang, H. Study on the Optimal Allocation of Public Service Facilities from the Perspective of Living Circle—A Case Study of Xiangyang High-Tech Zone, China. *Journal of Urban Management*, 2023, *12*, 344–359, doi:10.1016/j.jum.2023.07.004.
8. Wang, L.; Yu, Z.; Yang, D.; Ma, H.; Sheng, H. Efficiently Targeted Billboard Advertising Using Crowdsensing Vehicle Trajectory Data. *IEEE Transactions on Industrial Informatics*, 2020, *16*, 1058–1066, doi:10.1109/TII.2019.2891258.

9. Wang, Z.; Cao, K. An Intelligent Site Selection Approach for Public Service Facilities Coupled with Improved Graph Attention Network and Deep Reinforcement Learning[J]. *Journal of Geo-Information Science*, 2024, *26*(11), 2452–2464, doi:10.12082/dqxxkx.2024.240044.
10. Tong, D.; Murray, A.T. Spatial Optimization in Geography. *Annals of the Association of American Geographers*, 2012, *102*, 1290–1309, doi:10.1080/00045608.2012.685044.
11. Laporte, G.; Nickel, S.; Saldanha-da-Gama, F. Introduction to Location Science. In *Location Science*; Laporte, G., Nickel, S., Saldanha Da Gama, F., Eds.; Springer International Publishing: Cham, 2019; pp. 1–21.
12. Land, A.H.; Doig, A.G. An Automatic Method for Solving Discrete Programming Problems. In *50 Years of Integer Programming 1958–2008*; Jünger, M., Liebling, T.M., Naddef, D., Nemhauser, G.L., Pulleyblank, W.R., Reinelt, G., Rinaldi, G., Wolsey, L.A., Eds.; Springer Berlin Heidelberg: Berlin, Heidelberg, 2010; pp. 105–132.
13. Williamson, D.P.; Shmoys, D.B. *The Design of Approximation Algorithms*; Cambridge University Press, 2011.
14. Davidon, W.C. Variable Metric Method for Minimization. *SIAM Journal on Optimization*, 1991, *1*, 1–17, doi:10.1137/0801001.
15. Song, Z.; Yan, T.; Liu, T.; Huang, T. A New Gravity P-Median Model and Empirical Test in Urban Comprehensive Hospital Location Decision Making: Take Wuxi as an Example. *Progress in Geography*, 2016, *35*(4), 420–430.
16. Kong, Y.; Wang, X.; Wang, Z. Solving p-Median Problem with MIP Optimizers: A Case Study of School Site Selection. *Journal of Henan University (Natural Science)*, 2014, *44*(6), 725–730.
17. Huang, Y.; Wang, X.; Chen, H. Location Selection for Regional Logistics Center Based on Particle Swarm Optimization. *Sustainability*, 2022, *14*, 16409, doi:10.3390/su142416409.
18. Yang, E. Research on the Layout of Highway Transportation Hub Bus Stations Based on the P-Median Method (Master's thesis). Beijing Jiaotong University, 2009.
19. Mo, H.; Chen, Q.; Zhang, Z. Solving the Electric Power Facility Location Problem Using Genetic Algorithms. *Computer Technology and Development*, 2016, *26*(3), 197–201.
20. Bai, C. Semi-Obnoxious, Obnoxious p-Median Problems and Location Problems with Connectivity Constraints (Doctoral dissertation). Shanghai University, 2015.
21. Chen, Z.; Wang, W.; Li, Z.; Zhang, Y.; Sun, Z.; Sun, X. Optimization of Charging Pile Layout Based on Multi-Factor Constrained P-Median Model. *Journal of Beijing Jiaotong University*, 2021, *45*(3), 93–99.
22. Dogondaji, Z.; Ojonugwa, E.; Ibrahim, A.; Mustafa, A. P-Center Problem of Fire Stations in SOKOTO METROPOLIS.
23. Lu, C.-C.; Sheu, J.-B. Robust Vertex P-Center Model for Locating Urgent Relief Distribution Centers. *Computers & Operations Research*, 2013, *40*, 2128–2137, doi:10.1016/j.cor.2013.02.019.
24. Garfinkel, R.S.; Neebe, A.W.; Rao, M.R. The M-Center Problem: Minimax Facility Location. *Management Science*, 1977, *23*, 1133–1142, doi:10.1287/mnsc.23.10.1133.
25. Elloumi, S.; Labbé, M.; Pochet, Y. A New Formulation and Resolution Method for the P-Center Problem. *INFORMS Journal on Computing*, 2004, *16*, 84–94, doi:10.1287/ijoc.1030.0028.

The perceived complexity of urban environments arises from multiple interacting elements (Guan et al., 2022), including: visual features such as color, shape, texture. The complexity of urban built environment directly affects human perception and interpretation of urban environment (Kraff et al., 2020). Besides, structural characteristics such as the branching patterns of road networks, the density of intersections, and the layout of pathways shape the perceived structural complexity. In addition, semantic features, such as the number and spatial distribution of points of interest (POIs), as well as recognizable landmarks, add a layer of meaning to the visual and structural components.

Understanding how humans perceive and mentally organize urban environments is also an intricate task. Perceptual processes not only reflect people's spatial cognition, shape citywide navigation behaviors but also influence a broader range of urban dynamics, including mobility patterns, social interactions, and planning decisions. By modeling the perceived scene complexity of real-world environments—using tools like street-view panoramas and geospatial data—researchers can gain deeper insights into how urban elements impact human cognition and behaviors (Yao et al., 2021).

8.1.1 Understanding Human Spatial Cognition in Cities

Urban environments are inherently dense with semantic information, encompassing visual elements like landmarks, road structures, and traffic signs (Campbell et al., 2019). The study of human perception and spatial cognition in urban environments has gained significant traction with the advent of geographic big data. For instance, street-view imagery has been widely used to evaluate human perceptions of urban visual environments, offering a systematic approach to quantify factors such as visual richness and semantic complexity. Tools such as street-view panoramas and open geodata enable researchers to analyze urban environments across thousands of cities with high efficiency. By modeling and computing indicators of environmental complexity, these datasets—when combined with cognitive experiments—provide valuable insights into how humans perceive and mentally organize urban spaces.

Existing research reveal that urban spaces are not perceived in their entirety but are abstracted and distilled into key elements, including paths, edges, districts, nodes, and landmarks, as described in Kevin Lynch's seminal work, *The Image of the City*. These elements form the building blocks of human spatial cognition, facilitating navigation and shaping the "mental map" of the city (Lynch, 1964).

This perceptual process of urban environment involves a two-way interaction between the human observer and the surrounding space. As Kevin Lynch emphasized, environmental images are shaped by this interplay: the environment provides distinctions and relationships, while the observer selects and organizes these cues based on personal purposes and adaptability. Through this extraction and abstraction process, individuals are able to synthesize vast and complex urban information into coherent mental representations or cognitive maps, which are essential for navigation and spatial reasoning. These maps reflect how people simplify and internalize the complexity of urban spaces.

From individual perception to broader urban dynamics, it becomes clear that understanding human spatial cognition is vital for designing environments that accommodate collective human needs. For urban planners, they focus not only on the individual experience but also on achieving a shared mental image of the city among its inhabitants—a consensus that emerges from common experiences of landmarks, road networks, and districts, as well as the overall structure of the urban environment. This shared cognitive framework, often referred to as *The Image of the City*, fosters intuitive navigation, strengthens community identity, enhances the sense of belonging, and improves urban functionality (Filomena et al., 2019).

8.1.2 Cognitive Maps for Human Navigation Behaviors in Cities

Cognitive maps are essential for understanding human navigation behaviors in urban environments. These mental representations of the city allow individuals to simplify and organize complex spatial information, aiding in tasks such as wayfinding and navigation. Rather than perceiving cities as a whole, people abstract urban elements like landmarks, road structures, and districts into key components that form the foundation of their cognitive maps. These mental schemas are not only practical for navigation but also integral to how humans perceive and interact with their urban surroundings.

Visual cues play a crucial role during the formation of cognitive maps in human navigation behaviors, with landmarks playing a pivotal role by serving as visually distinct reference points that aid in spatial understanding and decision-making. Their importance extends beyond navigation to influence place identity, community belonging, and mental route representations (Bruns & Chamberlain, 2019). Studies have demonstrated that landmarks provide essential environmental support for establishing place identity and fostering a sense of belonging within geographic communities. Recognizable landmarks anchor individuals to specific locations, creating meaningful connections between physical spaces and personal or communal identities. This process not only simplifies navigation but also cultivates emotional and cultural ties to the city, enriching the overall urban experience of living and moving within a community.

Research has shown that landmarks hold a special cognitive status in mental route representations, as they reduce cognitive load during navigation. Compared to street name-based directions, landmark-based instructions significantly shorten the time needed for cognitive processing, making navigation quicker and more intuitive. This suggests that visual cues are more efficiently processed by the brain, offering more reliable navigation assistance. By using landmarks with distinct and memorable visual anchors, navigators can more easily encode routes, making them easier to recall and apply than abstract textual information such as street names (Tom & Denis, n.d.).

While visual landmarks are crucial for the formation of cognitive maps, human perception often involves multimodal inputs, such as auditory cues or tactile feedback, which complement visual information. These additional sensory signals enhance human's spatial cognition and adaptability, particularly in dynamic environments. The integration of various sensory inputs allows individuals to make decisions more effectively, enhancing their overall ability to orient themselves and perform navigation behaviors in complex urban spaces.

8.1.3 The Biological Foundation of Navigation

Spatial navigation process refers to an individual's ability to locate themselves, move, and remember different positions in the environment. This process involves the brain's handling and recalling of spatial information. It is intricately linked to specific brain regions that help navigate and form a cognitive map. Among these regions, the hippocampus plays a central role in encoding, storing, and retrieving spatial memory. Hippocampus functions as a dynamic map, capturing the individual's experience of a place and its associated events. In essence, it organizes spatial memories based on personal experiences and the contextual features of the environment (Deuker et al., 2016; Farzanfar et al., 2023). The existence of this structure provides a traceable understanding of navigation from a biological perspective.

Key biological foundation involved in navigation include place cells, grid cells, and border cells. Place cells, primarily located in the hippocampus, are activated when an individual is at a specific location in the environment (Moser et al., 2008). These cells encode salient locations and are crucial for identifying landmarks or points of interest. According to research by O'Keefe and Nadel (O'keefe & Nadel, 1979) and later studies, place cells form the core of the cognitive map, facilitating spatial memory by identifying crucial locations (Doeller et al., 2008). Similarly, grid cells, found in the entorhinal cortex, encode information about the distances and directions between different locations, providing a metric framework for navigation. These cells activate in a grid-like pattern across space, helping the brain calculate distances and navigate efficiently. Meanwhile, border cells are responsible for signaling the boundaries of a region, providing critical information about the edges of a navigable space. These biological cells in the brain collaborate to shape an individual's spatial perception, offering guidance for navigation.

As part of the above process, the spatial representation of the environment in the brain evolves from egocentric (self-centered) to allocentric (world-centered) maps. The initial egocentric map consists of a set of observed features aligned along a route, such as specific landmarks or sensory cues, with respect to the individual's own position. As one's experience with a particular space increases, these features become integrated into a more sophisticated allocentric structure that captures the relationships between features independent of the individual's current position. This transition from an egocentric to an allocentric map is essential for advanced navigation, allowing individuals to navigate unfamiliar environments and perform spatial tasks with a broader understanding of their surroundings.

Navigation tasks, especially in urban environments, provide insights into the neural foundations of cognitive maps. Recent studies highlight how neural activity encodes spatial topology and reacts dynamically to the structural characteristics of cityscapes. Research has shown that the population activity of neurons forms a neural manifold in high-dimensional space, reflecting spatial representations during navigation tasks (Yang et al., 2024). By applying dimensionality reduction methods, the topology of environments, such as mazes, can be explicitly mapped onto a low-dimensional manifold. These neural manifolds represent the spatial layout of the navigation task, providing direct evidence of how cognitive maps emerge as organized neural patterns in the brain. This process of neural encoding and abstraction provides a foundation for understanding how the brain processes spatial information in complex environments, such as urban areas. It demonstrates that neural activity not only encodes spatial structure but also enables efficient mental navigation through abstracted spatial representations.

Further research reveals that spatial topological characteristics of urban environments, such as the degree centrality of a street segment in the road network, correlate with brain activity during navigation (Javadi et al., 2017), in which degree centrality refers to the importance of a street segment based on the number of connections it has to other segments. Functional MRI studies reveal that, following training in navigating unfamiliar districts, changes in the degree centrality of road segments during navigation are specifically correlated with increased activity in the right posterior hippocampus. This effect is most pronounced when participants enter highly central street segments, indicating that urban environments inherently contain cognitive elements that align with the brain's spatial processing mechanisms. Such findings suggest that the brain dynamically adapts its spatial representations to reflect the structural hierarchy of cityscapes.

8.2 Perceiving Cities through Human Navigation

8.2.1 Human Navigation Behaviors Can Reveal Urban Built Environment

Human navigation behaviors can reveal much about the urban built environment, offering insights into how the design and structure of cities influence the way people move and interact with space. Urban navigation is often conceptualized as a utility optimization problem, where decision-making involves balancing several factors, not merely minimizing the physical distance. For example, a vector-based pedestrian navigation model suggests that people frequently prioritize mental burden minimization over the shortest path (Bongiorno et al., 2021). This means that pedestrians tend to prefer straightforward, cognitively simpler routes over more complex, winding paths, even if the latter may be shorter in distance. This behavioral pattern underscores the psychological and cognitive aspects involved in route choice, and importantly, highlights how urban design influences such decisions.

Human navigation behaviors are deeply influenced by the urban built environment, revealing how urban design and planning shape movement patterns and route choices within cities. Researchers have shown that the spatial characteristics of urban environments, including road networks, green spaces, and urban amenities, significantly impact decision-making in navigation. Pedestrians often choose routes that minimize mental and physical burden, even if those paths are longer, preferring areas with aesthetic and functional benefits such as greenery, parks, and urban furniture (Guo & Loo, 2013; Miranda et al., 2021). These preferences highlight the importance of environmental features in fostering desirable spaces, influencing the trajectory of movement and social interactions within urban areas. Furthermore, urban planning plays a pivotal role in these behaviors; for instance, high traffic volumes and a lack of pedestrian-friendly infrastructure can discourage movement, especially for vulnerable groups like the elderly (Borst et al., 2009). In contrast, streets with gardens, residential zones, and shops are perceived as safer and more accessible, attracting more pedestrians. Such findings underscore how urban design can shape the cognitive and emotional experiences of individuals, offering insights into how urban planning decisions can foster more efficient, safe, and appealing environments for city dwellers.

Moreover, Inverse Reinforcement Learning (IRL) offers a powerful tool for modeling urban navigation by analyzing historical route data (Ziebart et al., n.d.). By observing human movements, IRL can derive policies that predict optimal routes, destinations, and subsequent steps within a city, effectively capturing the implicit decision-making strategies of urban navigators, allowing us to account for both environmental cues and individual preferences.

8.2.2 Cognitive Maps of Cities Learned by AI Navigation

Although existing research has quantitatively analyzed the relationship between navigation behavior and certain environmental factors within cities using real human navigation trajectory data, it is challenging to conduct large-scale experiments with pedestrian trajectory data across different cities. This limitation makes it difficult to expand these studies beyond specific urban contexts. In addition, visual observation is critical for navigators in making decisions, yet traditional human navigation studies often rely on statistical analysis of route choice and environmental factors. AI-powered navigators, however, can simulate human perception of the city, enabling navigation behavior decisions at a broader scale. This presents a promising way to address the challenges of existing research. Moreover, through advancements in predictive modeling and deep reinforcement learning, current AI research has already demonstrated, to some extent, that just as human brains represent urban space through cognitive maps, AI systems that acquire navigation abilities in virtual and real-world settings can also develop their own cognitive maps within their representations based on urban observations.

In a virtual geographical environment, an agent was trained using visual predictive coding without access to explicit environmental coordinates or maps (Gornet & Thomson, 2024). This agent learns to anticipate future observable scenes and derives a cognitive map from the neural network's final layer representation. Remarkably, the representation exhibits biologically similar traits, such as place cell-like activity, where artificial neurons are selectively activated in specific regions. Furthermore, the discovery of representations supporting vector-based navigation mechanisms highlights parallels between AI predictive coding and biological navigation systems. Another significant advancement is the application of a deep reinforcement learning approach for city-scale navigation, also without access to explicit city maps (Mirowski et al., 2018). Using Google Street View as a simulation environment, researchers trained agents to integrate general policies with locale-specific knowledge, enabling robust long-range navigation across multiple cities. The process of learning navigation through reinforcement learning is very similar to human memory and exploration of a spatial area. By processing visual sensory input and receiving information about the target location, the system makes decisions on movement behavior and receives reward signals based on performance. Through trial and error and exploration, it learns navigation abilities. Intriguingly, the internal representations developed by these agents allow for the decoding of their positions, functioning as a cognitive map for navigation. This suggests that AI not only learns efficient strategies for reaching destinations but also builds an internalized, map-like representation of its environment. Collectively, these findings show that AI can autonomously construct cognitive maps, bridging the gap between biological and artificial representations of space, and paving the way for more adaptive, human-like navigation systems.

In the future, these AI navigators could explore the formation of spatial memory and its transformation as they navigate between new and old cities and spatial structures, further aligning with concepts like "spatial schema" from neuroscience.

8.2.3 Embodied Urban Navigation for Perceiving Cities

The integration of various AI techniques significantly enhances understanding human perception to cities by providing diverse approaches to understanding and interacting with complex environments in an embodied perspective. Deep reinforcement learning offers a pathway to embodied perception by allowing agents to learn navigation strategies through direct interaction with their surroundings. This enables agents to adapt to the dynamics of urban spaces and develop spatial awareness based on experience. In addition, transfer learning accelerates the adaptation of agents across different cities by leveraging knowledge gained in one urban context and applying it to another, facilitating more efficient learning in new environments. Self-supervised learning might further complement these approaches by allowing agents to build robust representations of the environment without the need for labeled data, enhancing the agent's ability to navigate autonomously. Moreover, large language models (LLMs) have emerged as powerful tools for interpreting urban contexts, enriching spatial understanding by integrating textual data such as geographic descriptions, urban policies, and other related information. Recent research suggests that simulating the "Mind's Eye" process of human spatial reasoning can enhance the spatial understanding ability of LLMs by visualizing its thinking process during executing navigation tasks (Wu et al., 2024). The navigation capability of LLM-based applications demonstrated by the existing research shows application value for urban development. LiMeda (Chen et al., 2024), a multi-vehicle scheduling and navigation framework driven by LLM, was proposed to solve the problem of incompatible vehicle resources in smart city scenarios and assist multiple types of vehicles in completing collaborative navigation tasks in cities. It can be foreseen that in the field of urban navigation in the future, LLMs will further explore urban information in spatial cognition and inference decision-making, promote social development and rational allocation of resources. These advances, when combined, offer a more holistic view of human perception to cities, combining visual, contextual, and experiential insights into the fabric of city environments.

Representation learning is increasingly significant for understanding human perception to cities. Large-scale pretrained models offer great efficiency in representations, but may lack the human-like nuances captured during navigation tasks. Reinforcement learning (RL) based navigation agents can also develop representations of city space during training, capturing the perspective of human pedestrians and the complexities of real-time decision-making. However, RL methods can be computationally intensive and less efficient compared to self-supervised models. Future research could investigate hybrid approaches, such as integrating transfer learning with offline RL, to enhance both efficiency and robustness in representation learning using reinforcement learning. The trade-off between these approaches, as well as the balance between efficiency and meaningfulness, will be an important area for further exploration.

References

Bongiorno, C., Zhou, Y., Kryven, M., Theurel, D., Rizzo, A., Santi, P., Tenenbaum, J., & Ratti, C. (2021). Vector-based pedestrian navigation in cities. *Nature Computational Science, 1*(10), 678–685. https://doi.org/10.1038/s43588-021-00130-y

Borst, H. C., de Vries, S. I., Graham, J. M., van Dongen, J. E., Bakker, I., & Miedema, H. M. (2009). Influence of environmental street characteristics on walking route choice of elderly people. *Journal of Environmental Psychology, 29*(4), 477–484.

Bruns, C. R., & Chamberlain, B. C. (2019). The influence of landmarks and urban form on cognitive maps using virtual reality. *Landscape and Urban Planning, 189*, 296–306.

Campbell, A., Both, A., & Sun, Q. C. (2019). Detecting and mapping traffic signs from google street view images using deep learning and GIS. *Computers, Environment and Urban Systems, 77*, 101350.

Chen, R., Song, W., Zu, W., Dong, Z., Guo, Z., Sun, F., Tian, Z., & Wang, J. (2024). An LLM-driven framework for multiple-vehicle dispatching and navigation in smart city landscapes. *2024 IEEE International Conference on Robotics and Automation (ICRA)*, 2147–2153. https://ieeexplore.ieee.org/abstract/document/10610578/

Deuker, L., Bellmund, J. L., Navarro Schröder, T., & Doeller, C. F. (2016). An event map of memory space in the hippocampus. *eLife, 5*, e16534. https://doi.org/10.7554/eLife.16534

Doeller, C. F., King, J. A., & Burgess, N. (2008). Parallel striatal and hippocampal systems for landmarks and boundaries in spatial memory. *Proceedings of the National Academy of Sciences, 105*(15), 5915–5920. https://doi.org/10.1073/pnas.0801489105

Farzanfar, D., Spiers, H. J., Moscovitch, M., & Rosenbaum, R. S. (2023). From cognitive maps to spatial schemas. *Nature Reviews Neuroscience, 24*(2), 63–79.

Filomena, G., Verstegen, J. A., & Manley, E. (2019). A computational approach to 'The image of the city.' *Cities, 89*, 14–25. https://doi.org/10.1016/j.cities.2019.01.006

Gornet, J., & Thomson, M. (2024). Automated construction of cognitive maps with visual predictive coding. *Nature Machine Intelligence*. https://doi.org/10.1038/s42256-024-00863-1

Guan, F., Fang, Z., Wang, L., Zhang, X., Zhong, H., & Huang, H. (2022). Modelling people's perceived scene complexity of real-world environments using street-view panoramas and open geodata. *ISPRS Journal of Photogrammetry and Remote Sensing, 186*, 315–331. https://doi.org/10.1016/j.isprsjprs.2022.02.012

Guo, Z., & Loo, B. P. (2013). Pedestrian environment and route choice: Evidence from New York City and Hong Kong. *Journal of Transport Geography, 28*, 124–136.

Javadi, A.-H., Emo, B., Howard, L. R., Zisch, F. E., Yu, Y., Knight, R., Pinelo Silva, J., & Spiers, H. J. (2017). Hippocampal and prefrontal processing of network topology to simulate the future. *Nature Communications, 8*(1), 14652. https://doi.org/10.1038/ncomms14652

Kraff, N. J., Wurm, M., & Taubenböck, H. (2020). Uncertainties of human perception in visual image interpretation in complex urban environments. *IEEE Journal of Selected Topics in Applied Earth Observations and Remote Sensing, 13*, 4229–4241.

Lynch, K. (1964). *The image of the city*. MIT press. https://books.google.com/books?hl=zh-CN&lr=&id=_phRPWsSpAgC&oi=fnd&pg=PA1&dq=kevin+lynch+image+of+city&ots=jJB59a2Dhl&sig=4QYqtLXiRebnZKqJ2CsaZkB2axQ

Miranda, A. S., Fan, Z., Duarte, F., & Ratti, C. (2021). Desirable streets: Using deviations in pedestrian trajectories to measure the value of the built environment. *Computers, Environment and Urban Systems, 86*, 101563.
Mirowski, P., Grimes, M., Malinowski, M., Hermann, K. M., Anderson, K., Teplyashin, D., Simonyan, K., Zisserman, A., & Hadsell, R. (2018). Learning to navigate in cities without a map. *Advances in Neural Information Processing Systems, 31*. https://proceedings.neurips.cc/paper/7509-learning-to-navigate-in-cities-without-a-map
Moser, E. I., Kropff, E., & Moser, M.-B. (2008). Place cells, grid cells, and the brain's spatial representation system. *Annual Review of Neuroscience, 31*(1), 69–89. https://doi.org/10.1146/annurev.neuro.31.061307.090723
O'keefe, J., & Nadel, L. (1979). Précis of O'Keefe & Nadel's The hippocampus as a cognitive map. *Behavioral and Brain Sciences, 2*(4), 487–494.
Tom, A., & Denis, M. (2004). Language and spatial cognition: Comparing the roles of landmarks and street names in route instructions. Applied Cognitive Psychology: The Official Journal of the Society for Applied Research in Memory and Cognition, 18(9), 1213–1230.
Wu, W., Mao, S., Zhang, Y., Xia, Y., Dong, L., Cui, L., & Wei, F. (2024). Mind's eye of LLMs: Visualization-of-thought elicits spatial reasoning in large language models. *The Thirty-Eighth Annual Conference on Neural Information Processing Systems*. https://openreview.net/forum?id=CEJ1mYPgWw
Yang, W., Sun, C., Huszár, R., Hainmueller, T., Kiselev, K., & Buzsáki, G. (2024). Selection of experience for memory by hippocampal sharp wave ripples. *Science, 383*(6690), 1478–1483. https://doi.org/10.1126/science.adk8261
Yao, Y., Wang, J., Hong, Y., Qian, C., Guan, Q., Liang, X., Dai, L., & Zhang, J. (2021). Discovering the homogeneous geographic domain of human perceptions from street view images. *Landscape and Urban Planning, 212*, 104125.
Ziebart, B. D., Maas, A., Bagnell, J. A., & Dey, A. K. (n.d.). *Maximum Entropy Inverse Reinforcement Learning* , Vol. 8, pp. 1433-1438.

9

Toward the Science of Living Structure: Making and Remaking Livable Cities as Part of Urban Informatics

Bin Jiang, Qianxiang Yao, Huan Qian, and Bisong Hu

9.1 Introduction

Contemporary fields of architectural and urban design are going through a major transformation, which has been driven by growing acknowledgment of the limitations and failures of modernist approaches. This transformation was highlighted by an open letter (Architectural Education Declares 2019) that over 2500 students and professionals signed, which criticized the lack of sensitivity that current methodologies have regarding environmental and social contexts and called for a new paradigm emphasizing human well-being, sustainability, and a harmonious relationship with nature. This institutional critique runs parallel to a grassroots movement called the Architectural Uprising (2014), which supports a return to traditional architectural principles, with an emphasis on local materials, craftsmanship, and vernacular styles. Advocates of the Architectural Uprising argue that such principles are more suitable for the creation of beautiful, livable, and enduring urban environments. They call for designs that resonate with local culture and human experience, rejecting what they believe is the often unsustainable, cold, and impersonal nature of modernist architecture. These calls are supported by the Chernobyl Paradox (Mehaffy and Salingaros 2020), which highlights the long-term consequences of poorly conceived architectural and urban interventions and warns of the dangers of overlooking the interplay between human settlements and their environments.

In the context of urban design, Jacobs (1961) claimed that a lot of modern urban planning was based on nonsensical principles that do not account for cities' organic and intricate nature. Jacobs challenged the mechanistic views of urban spaces, instead calling for cities to be understood as dynamic, living systems. More recent critics of contemporary urban design have included Cuthbert (2007) and Marshall (2012), who argued that urban design practices often lack solid scientific foundations and rely on oversimplified or outdated models that do not capture the complex nature of actual urban environments.

DOI: 10.1201/9781003484219-9

Together, such movements and critiques underline the need for architecture and urban design to create a new conceptual framework that embraces sustainability, complexity, and human-centered principles. The present paper introduces the concept of living structure as a transformative paradigm that attends to this need. Understanding and designing urban environments as living structures makes it possible to create spaces that are functional and also resonant with human experiences and ecological realities.

The science of living structure dramatically changes our understanding of environmental and spatial complexity. A central aspect of this science is the theory of centers (Alexander 2002–2005), which is also known as substructures and is supported by 15 fundamental properties (see Section 9.2) that define the intrinsic vitality and beauty of a space. This theory is operationalized through innovative methods, such as mathematical models that compute the degree of livingness (Salingaros 1997; Jiang 2015a; Jiang and de Rijke 2023), and the mirror-of-the-self experiment (Alexander 2002–2005), which uses humans and their biometric data obtained from functional near-infrared spectroscopy (fNIRS), eye-tracking, functional magnetic resonance imaging (fMRI), and electroencephalograms (EEG) as measuring instruments with which to gauge the psychological and emotional impact of a space (Sussman 2014, Salingaros and Sussman 2020). Living structure leverages open-access geospatial big data to analyze and enhance urban environments through the integration of such diverse sources as OpenStreetMap, nighttime light imagery, and social media data. This multifaceted approach makes it possible to better understand and create spaces that are not functional, profoundly beautiful, and life-enhancing.

The remainder of this paper is structured as follows. Section 9.2 examines the concept of living structure and its 15 properties. Section 9.3 investigates structural beauty and the vitality of space, as quantified by the formula $L = S \times H$, where L represents livingness, S indicates the number of substructures, and H is their inherent hierarchy. Section 9.4 presents three illustrative case studies – Alexander's 253 patterns, two pairs of satellite images, and a design example – show the application of the principles of living structure. Section 9.5 looks at the science of living structure, proposes a shift from a mechanistic to an organic view of space, and outlines a research agenda. Section 9.6 summarizes the main points and implications.

9.2 Living Structure and Its Geometrical and Transformation Properties

Christopher Alexander's theory of living structure forms the foundation of the science of living structure, identifying 15 fundamental properties that are universally present in living structures. These properties, which include levels of scale, strong centers, alternating repetition, thick

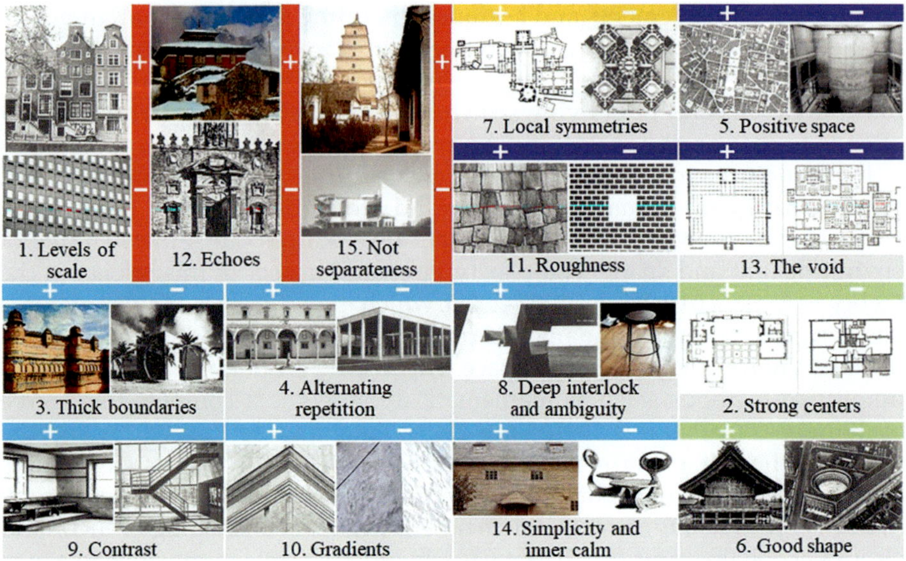

FIGURE 9.1
(Color online) The 15 geometrical and transformation properties of living structure.

Note: Each property is represented by positive (+) and negative (–) examples, primarily from Alexander (2002–2005). The examples are placed into five categories, each marked by a different color. Red indicates global properties that span different scales, while the properties in the two blue groups are local. The 'local symmetries' property is highlighted separately because of its unique, pre-existing discovery prior to the set of the 15 properties. 'Strong centers' and 'good shape' are considered to be redundant because they are essentially living structures themselves.

boundaries, and positive space (Figure 9.1), capture the essence of what makes a structure beautiful and living, and serve as guidelines for designing engaging, coherent, and vibrant spaces. Adherence to these properties can enable a designer to create an environment that fosters a sense of well-being and attachment.

The first of these properties, and the most important, is the levels of scale, which makes sure that a structure has multiple levels of scale, ranging from small details to large overarching forms. This property can be formulated as the scaling law, which is the recurring notion of far more smalls than larges (Jiang 2015b). This concept, which is also known as scale jump (Alexander 2002–2005), will be discussed further in the next section, with a description of structural beauty or the liveliness of space. The second property is strong centers, which refers to the creation of strong and coherent centers that organize the surrounding elements (or, more precisely, centers or substructures). This property is actually redundant, given that all of the other properties are intended to create strong centers, probably with an exception of the first and the last properties. The third property is thick boundaries, which defines the characteristics of boundaries that distinguish different areas and elements (or substructures in general) within the structure. Next comes alternating repetition, which refers to the use of alternating patterns to create movement and rhythm within the space; this contrasts with the monotonous repletion that is often apparent in modernist buildings and cities. The next property is positive space, which ensures that both the figure and the ground are well-shaped and contribute to the overall form. The sixth property is good shape, which is best understood in a recursive manner, as meaning that a good shape consists of numerous good shapes. According to the recursive definition, this property of good shape aligns with the living structure itself and is therefore considered to be redundant. The seventh property, local symmetries, incorporates symmetrical elements to create balance and harmony and mainly refers to symmetries at local levels rather than on the global levels.

The eighth property is *deep interlock and ambiguity,* which creates complex relationships between adjacent substructures that enhance the depth and richness of a space. Next is *contrast,* which means using contrast to highlight differences and create visual interest, followed by *gradients,* which means shoring or creating gradual changes in color, form, or texture to create a sense of flow and continuity. The 11th property is *roughness;* that is, incorporating variations and irregularities in order to avoid monotony and to create a natural appearance. The 12th property, *echoes,* means repeating substructures to create a sense of coherence and unity at different levels of scale. Property number 13 is *the void,* which refers to the inclusion of open spaces that provide contrast and enhance the surrounding substructures. Next is *simplicity and inner calm,* which means striving for simplicity and calmness in some substructures at the same time as maintaining complexity across various levels of scale. Finally, the 15th property is *not separateness,* which ensures that substructures are well integrated and not isolated from one another.

A structure that exhibits these 15 properties to a high degree is considered vibrant and beautiful. However, these properties should be applied holistically, not in isolation. The three types of substructures that need to be considered in any given space are smaller spaces within it, adjacent spaces around it, and the larger space that contains it. Alexander (2002–2005) stated that a space is deemed to be 'living' or 'alive' if it satisfies all three of these criteria, meaning that a space is living when its adjacent spaces, the larger containing space, and the smaller spaces within it are all living. These 15 properties can be further distilled into two fundamental laws: the scaling law (Jiang 2015b), which posits that across different scales, there are significantly more small substructures than large ones; and Tobler's law (Tobler 1970), which states that substructures at any level of scale are more or less similar to one another.

9.3 Structural Beauty or Spatial Vitality

The concept of structural beauty – or the livingness of space (L) or spatial vitality – is a critical aspect of living structure that underlines the intricate relationship between the number of substructures (S) within a space and the hierarchical arrangement of those substructures (H). This relationship is expressed mathematically as $L = S \times H$. In order to understand and measure a space's livingness, it is necessary to look at its organic and hierarchical properties. From an organic perspective, space is envisioned in a similar way to biological cells that undergo transformation. Cells divide and form many substructures at various hierarchical levels, and space evolves in the same way. This perspective is in keeping with the principle that a space or subspace has a certain degree of livingness when it embodies far more small substructures than large ones. This recurring notion, which is also known as the levels of scale, hierarchy, or scale jumps, reiterates the essence of complexity within a living structure.

The measure of livingness in a space is tied to its complexity, which follows specific principles and patterns. The two key parameters that define this complexity are the number of substructures (S) and hierarchical arrangement (H). The former refers to the number of distinct, identifiable subspaces within a larger space, which can range from minute details to larger and more prominent features. When the number of substructures is higher, the livingness of the space is richer. H denotes the organization of substructures into different levels or scales. In this context, hierarchy means that smaller substructures are nested within larger ones, which creates a multilayered structure. As this hierarchy becomes more intricate and well-organized, the livingness increases. Together, the two parameters – S and H – contribute to the overall livingness (L) of a space. The interaction between many small substructures and their hierarchical arrangement creates a dynamic and engaging environment that reflects the principles of living structure.

The notions of scale jumps, hierarchy, levels of scale, and the recurring concept of having far more smalls than larges all describe the same fundamental idea, which is the presence of a complex and multilevel structure in which small elements are embedded within larger ones. This principle can be seen in both natural and human-made (Simon 1962, Bak 1996), ranging from the branching patterns of trees to the arrangement of a traditional urban environment. Figure 9.2 illustrates this concept further by comparing two buildings. The image on the left-hand side represents a traditional, organic building that is a landmark at Sun Yat-Sen University in Guangzhou, China, while the image on the right-hand side is a modernist building, Villa Savoye in Poissy, France, designed by Le Corbusier (1923), widely acknowledged as the father of modernist architecture. The traditional building has a livingness of $L(\text{left}) = (3 + 6 + 13 + 51 + 109 + 404) \times 6 = 3516$, while the

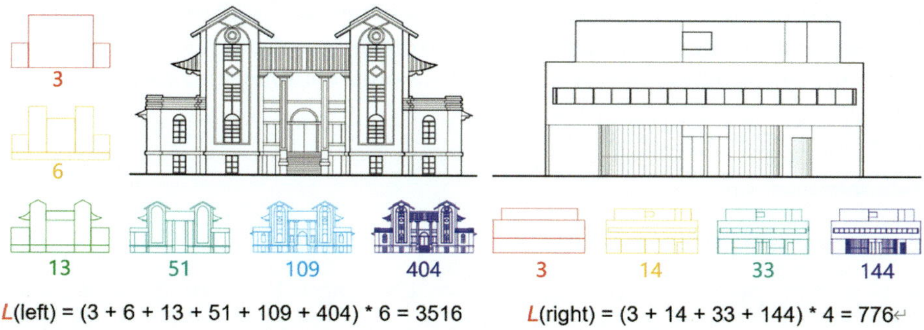

FIGURE 9.2
(Color online) The building on the left is more living than that on the right.

livingness of the modernist one is $L(\text{right}) = (3 + 14 + 33 + 144) \times 4 = 776$. The marked difference between the two buildings emphasizes how traditional buildings, which have many small and nested substructures, have considerably higher livingness than modernist buildings.

The concept of livingness or vitality has significant implications for architecture and urban planning. Designing spaces that adhere to the principles of living structure will enable architects and planners to create environments that are aesthetically pleasing and also promote well-being and a sense of place (Tuan 1977; Lewicka 2011; Seamon 2018). Rich substructures and their hierarchical organization can change ordinary spaces into vibrant and engaging environments. For instance, a traditional urban setting will often exhibit a high level of livingness due to its intricate street patterns, layered public spaces, and diverse building forms. By contrast, a modernist design that emphasizes uniformity and simplicity may lack such complexity, which will result in less engaging environments. The livingness of a space reflects its complexity and structural beauty. Focusing on the number of substructures and their hierarchical arrangement enables us to better understand and enhance the livingness of built environments. Such an approach aligns with scientific principles and resonates with humans' innate preference for complex and richly structured spaces.

The two fundamental design principles for creating living structures – differentiation and adaptation – align with the two abovementioned core laws of living structures. Differentiation is the process of transforming space through cell-like division, which generates multiple substructures across various scales, ensuring that each structure is unique and contributes to the richness of the overall form. Adaptation makes sure that all generated substructures harmonize with each other; while substructures can vary in terms of size and form, they require a level of coherence to ensure that each element relates to the structures surrounding it in such a way that preserves overall unity. While many other properties focus on achieving adaptation, the property of levels of scale mainly supports differentiation. These two principles – differentiation and adaptation – have been distilled from the 15 fundamental properties of living structure, which reflects the vital role they play in the production of spaces that are at once complex and coherent.

9.4 Illustrative Case Studies

This section presents three case studies that show the principles of living structure in different contexts. Each of the case studies applies hierarchical organization and the concept of "far more smalls than larges" to show the structural beauty and complexity of the subjects. The first case study looks at Alexander's 253 patterns and highlights their coherent network of substructures. The second study compares nighttime satellite images in order to assess livingness in both urban and rural areas. In the third case study, we detail the transformation of a classroom, highlighting hierarchical differentiation in creating engaging spaces. Together, these case studies showcase impact of practical applications of living structure theory in urban planning and design on various environments.

9.4.1 The 253 Patterns

The concept of living structure can be traced back to the 253 patterns, defined by Alexander et al. (1977), that addresses a range of problems and solutions. Each pattern is connected to both larger and smaller patterns, which creates a network like that of substructures. This case study investigates how these patterns form a coherent whole and computes their degrees of order or complexity. We organized the 253 patterns into six hierarchical levels, ranging from the largest (world government) to the smallest (daily life objects), which reveals a clear scaling hierarchy with far more small patterns than large ones. We then constructed a topological network based on the initial definitions of pattern associations. In contrast to previous studies, we arranged the patterns hierarchically and only allowed links within the same level or between adjacent levels, to ensure a clear hierarchical structure.

In this case study, we treated patterns as substructures that form a living structure. This is shown in Figure 9.3 in a spiral-shaped layout, which starts with the largest patterns and evolves to the smallest patterns, in the lowest hierarchical level, which exhibit the highest degrees of order or structural beauty; this reflects the principle that stronger differentiation leads to higher order. Each new generation of patterns emerges from the previous generation, inheriting and increasing the degree of order or complexity. This stepwise emergence can be seen in Figure 9.3, where each successive level is more orderly or more complex than the previous one. The 253-node network shows that the degree of order correlates with the recurring notion of far more smalls than larges, which makes the network visually appealing because it embodies structural beauty independent of color.

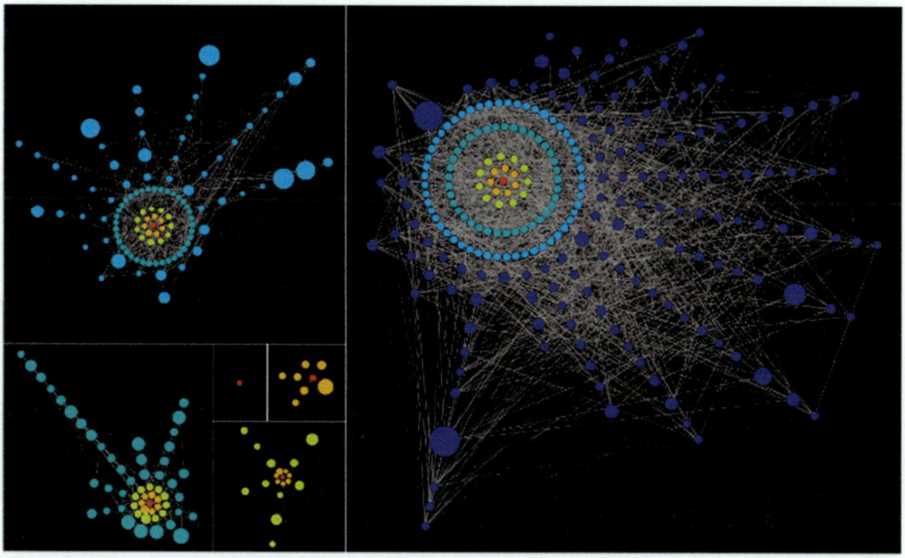

FIGURE 9.3
The notion of far more smalls than larges recurs multiple times.

Note: The layout follows a spiral shape to reflect the natural hierarchy within the 253 patterns. These patterns, as a coherent whole, emerge gradually from a single starting point – the idea of a world government – evolving step–by-step. Interestingly, the smaller patterns tend to show a greater degree of order or complexity, which highlights the key design principles of differentiation.

This case study shows how the 253 patterns form a coherent living structure. The hierarchical arrangement and the emergence of patterns at different scales reflect organized complexity. Such an approach quantifies the degree of order and also demonstrates how adaptation and differentiation principles guide the creation of living structures. This process mirrors natural human settlement patterns and emphasizes the importance of maintaining such principles in modern urban planning to avoid the creation of nonliving structures.

9.4.2 The Two Pairs of Satellite Images

The second case study involved comparing nighttime images of the Benelux (Belgium, the Netherlands, and Luxembourg) region with the Stockholm region (Figure 9.4, Panels a1 and b1). Because Benelux is more densely populated than Stockholm, it features a higher number of small human settlements, which translates to greater livingness, as reflected in the livingness scores (*LR* and *V*) and visualized in Panels a2, b2, a3, and b3 of Figure 9.4. The high density of small settlements in the Benelux region (Panel a1) creates complex, vibrant living structure, as can be seen in Panels a2 and a3 in the many first-iteration substructures (shown in blue), several intermediate

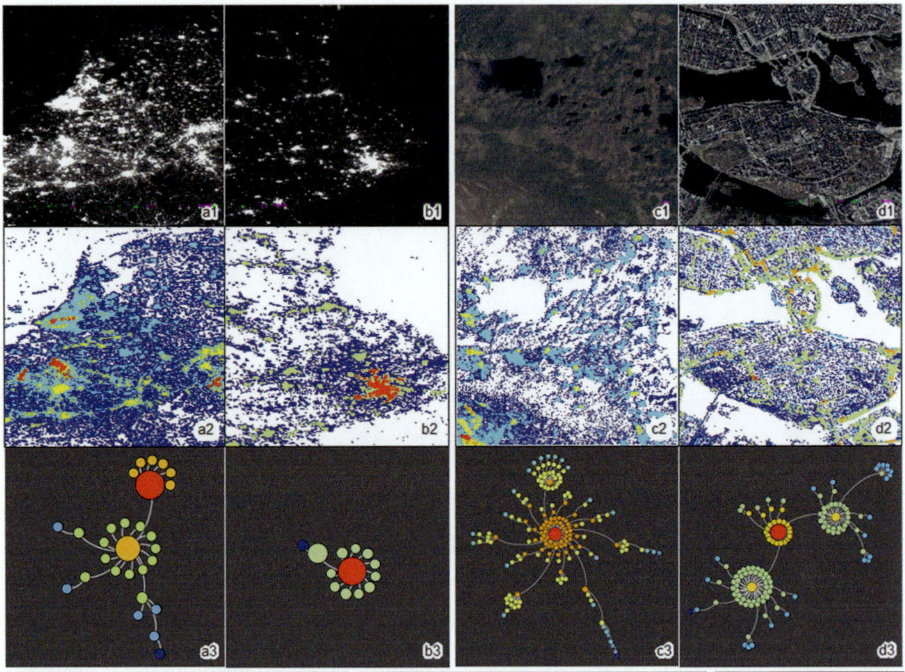

Nighttime Imagery	Living structure scores (a2, b2, a3, b3) S	H	LR	D	I	V
Benelux (a1)	2,974	4	11,896	15	5	75
Stockholm region (b1)	1,825	5	9,125	14	3	42

Orthophoto	Living structure scores (c2, d2, c3, d3) S	H	LR	D	I	V
Nature (Sorsele) (c1)	12,987	6	77,922	175	5	875
Urban (Stockholm) (d1)	8,913	5	44,565	131	5	655

Details	Nighttime	Orthophoto
Resolution	0.0083°	8m
Raster size	512*512	512*512
Source	EOS VIIRS	Lantmäteriet

FIGURE 9.4
(Color online) The recursive approach applied to the four georeferenced images.

Note: The source images (a1– d1); their corresponding living structures, both decomposable and undecomposable (a2– d2); and the decomposable living structures in graphical format (a3–d3). The living structures shown in the second row (a2–d2) have numerous first-iteration substructures (in blue), a few last-iteration substructures (in red), and some between the first- and last-iteration substructures (in colors on the spectrum between blue and red). Some of the substructures are nested to each other, with different colors showing the nested relationship, as shown in the second and third rows.

substructures (represented by various colors), and a few last-iteration substructures (red). By contrast, the Stockholm region (Panel b1) has fewer small settlements and exhibits lower livingness scores. Panels b2 and b3 display a less intricate hierarchical structure with fewer substructures.

The second pair of panels compares a satellite image of the Swedish countryside with the center of Stockholm city (Figure 9.4, Panels c1 and d1). While the two images have the same physical size and resolution, they capture different geographic features. The Swedish countryside (Panel c1) was expected to exhibit higher livingness because of its inherent complexity and variability, and this expectation was confirmed by the higher livingness scores (shown in Panels c2 and c3), displaying a rich structure with many first-iteration substructures (blue), intermediate substructures (various colors), and a few last-iteration substructures (red). The urban scene of Stockholm (Panel d1) shows lower livingness scores due to its structured and less complex layout, while Panels d2 and d3 reveal a simpler hierarchical structure with fewer substructures. The recursive decomposition of these images presents numerous first-iteration substructures (blue), a small number of last-iteration substructures (red), and intermediate substructures in various colors. Rows 2 and 3 of Figure 9.4 depict these nested relationships, highlighting each scene's hierarchical complexity.

The recursive approach (*LR* and *V*) captures the livingness of georeferenced images, showing that densely populated regions exhibit higher livingness scores than less populated regions, and natural scenes demonstrate higher livingness scores in comparison to urban areas. This method delivers a robust framework with which to quantify and visualize the livingness of different environments, highlighting the intricate structures that contribute to a region's overall sense of liveliness. The recursive approach, in revealing these underlying patterns, generates valuable insights into the spatial organization and complexity of certain geographic areas.

9.4.3 Renovation of a Classroom

Figure 9.5 shows the transformation of a classroom – through a careful design process that involves a thoughtful approach to how the space evolves and differentiate – from a plain and almost empty space into a vibrant and lively room. Initially, the room was basic and only contained essential furniture (Panel a). The walls were bare, and the overall feeling was sterile and uninviting. This minimal setup provided a 'blank canvas' that highlighted the potential for

FIGURE 9.5
(Color online) A small design process using the 15 transformation properties and in particular the levels of scale.

enhancement and the importance of thoughtful design to create a more engaging environment. Panel b shows the planning phase, in which different levels of scale were considered, including decisions regarding the number and size of the pictures to be hung. The table in Panel b shows three levels of scale – 1 (large), 1/3 (medium), and 1/9 (small) – with a corresponding number of paintings at each scale (one large, four medium, and 19 small). This careful consideration means that the elements are distributed in a balanced and aesthetically pleasing manner. This step is crucial because it lays the foundations for a harmonious design that emphasizes proportion and variety in visual interest.

Panel c shows the room starting to come to life. Pictures with varying sizes are placed strategically on the walls following the planned levels of scale, creating an engaging environment. The room feels warmer and more inviting than it did initially, while the artwork adds character and interest. The intentional placement of each picture shows how layering and including diverse elements can make a space more dynamic and stimulating. Panel d shows the room reaching its fully developed state. The design includes the pictures on the walls and integrates substructures that consider both the figure (the pictures) and the ground (the walls and the overall space). The room feels cohesive and harmonious, with the various substructures acting in conjunction to create a stimulating and lively environment. This final transformation shows the culmination of the design process, with all of the substructures contributing to a unified aesthetic that enhances the feel and function of the space. This small design process shows how paying careful attention to the arrangement and scale of the various substructures can turn a plain room into a vibrant and welcoming space. This approach reinforces the importance of design when creating functional and enjoyable environments.

Figure 9.6 shows a wall and its detailed substructures, particularly the inherent hierarchy that contributes to the wall being classified as a living structure with a very high degree of livingness. The wall has multiple layers, with

FIGURE 9.6
Substructures defined at different levels of scale for a wall of the classroom.

each having its own level of complexity and scale. The substructures are organized hierarchically, ranging from the largest elements (labeled with the number 3) to progressively smaller and more detailed components (ending with the number 1314). This hierarchy reflects how natural organisms are built, with larger systems encompassing smaller, detailed subsystems. Each substructure is interconnected with others to form a coherent whole, ensuring that changes in one part of the structure influence all the other parts and creating a responsive and dynamic system. For example, the largest substructures provide the main framework, while the smaller elements add detail and refinement, which enhances the overall aesthetic and functional qualities.

Using different levels of scale, with respective numbers of substructures—3, 13, 45, 145, 382, and 1314—demonstrates a thoughtful consideration of balance and proportion. Larger substructures provide the foundation, offering structure and stability, while smaller substructures add depth and intricacy, making the wall visually engaging and rich in detail, much like a living organism that can be observed at micro and macro levels. In addition to the levels of scale, the design incorporates multiple other essential characteristics of a living structure such as thick boundaries, local symmetry, and alternating repetition. This alternating repetition can be seen in the consistent use of panels, arches, and decorative motifs all across the wall. The wall's design process mimics organic growth, starting with the basic framework and progressively adding layers of detail. This approach is similar to the development of an embryo, with the entire structure giving rise to its parts through differentiation. The formula $L = S \times H = (3 + 13 + 45 + 145 + 382 + 1314) \times 6 = 11{,}412$ shows the cumulative complexity or livingness and the holistic nature of the design, with each part contributing to the living whole. By considering these substructures, the wall shifts from being a mere static structure to a living entity that is rich in detail, coherence, and organic complexity. This approach to design ensures that the space it inhabits is harmonious, engaging, and dynamic, embodying the principles of living structures found in nature.

9.5 Further Research on the Science of Living Structure

The science of living structure represents a shift from a mechanistic view of space to an organic one. The mechanistic view (Descartes 1644, 1983) perceives space as a compartmentalized and static entity that is focused on function and efficiency, which often leads to rigid and impersonal urban designs that are lacking in coherence and fail to engage users emotionally. The organic view (Whitehead 1929; Bohm 1980) considers space as a living system that is adaptive, interconnected, and flexible, emphasizing complexity, structural beauty, and human experience, with all elements interacting harmoniously with their surroundings. This shift is embodied in Alexander's 15 properties of living structure, all of which reflect the interconnectedness found in nature.

These properties are distilled into two key design principles: differentiation (the generation of distinct but related substructures) and adaptation, whereby the substructures fit together to create a unified whole. This organic approach helps create urban environments that are functional, aesthetically pleasing, and vibrant, and resonate with human experience. It is important to acknowledge the significance of two fundamental laws (the scaling law and Tobler's law) and two design principles (differentiation and adaptation) in shaping the fundamental framework ('skeleton') of living structures in architecture and urban design. Various architectural styles and cultural backgrounds could be regarded as the 'skin' that envelops the design process.

This article has illustrated several methodological case studies based on theoretical insights from the concept of living structure. Future research involves more empirical studies such as large-scale examples to validate the proposed theory and methodologies. Within the science of living structure, a critical research agenda is to leverage geospatial big data to reveal and analyze patterns of living structures in urban and architectural settings, which can enhance our understanding of human mobility, spatial dynamics, and the temporal evolution of urban areas. Advanced analytical techniques and computational models inspired by the notion of living structure enable researchers to reveal the underlying hierarchies and substructures that contribute to the livingness of a space. Such knowledge can inform urban planning and policy decisions, making cities more livable, sustainable, and resilient. Integrating geospatial big data, such as OpenStreetMap data, social media data, and nighttime imagery, makes it possible to study the dynamics of living structures and develop insights into how human activities shape and are shaped by their environments.

The concept of structural beauty extends beyond urban planning into a few other fields, including computer vision and image understanding. Applying the principles of living structure makes it possible to develop algorithms that assess the structural beauty of images, which can contribute in such areas as art, architectural design, and artificial intelligence (Jiang and de Rijke 2023). This interdisciplinary approach bridges the gap between functional and aesthetic considerations, fostering environments that are both visually appealing and conducive to human well-being, both physical and emotional. As various case studies have shown, the recursive approach to computing the structural beauty of images reveals that images that have higher levels of structural hierarchy and substructures are perceived as more beautiful than those with lower levels. Applying the same methodology to urban landscapes can help identify and enhance the livingness of cityscapes through improved design practices.

There is neuroscientific evidence that living structures have positive effects on human perception and cognition (e.g., Kawabata and Zeki 2004; Ishizu and Zeki 2011). Studies that utilize eye-tracking technology, fMRI, EEG, and fNIRS could show that environments with high structural beauty can reduce stress, enhance cognitive performance, and improve overall mental health (Figure 9.7). Furthermore, future research will also concentrate on the

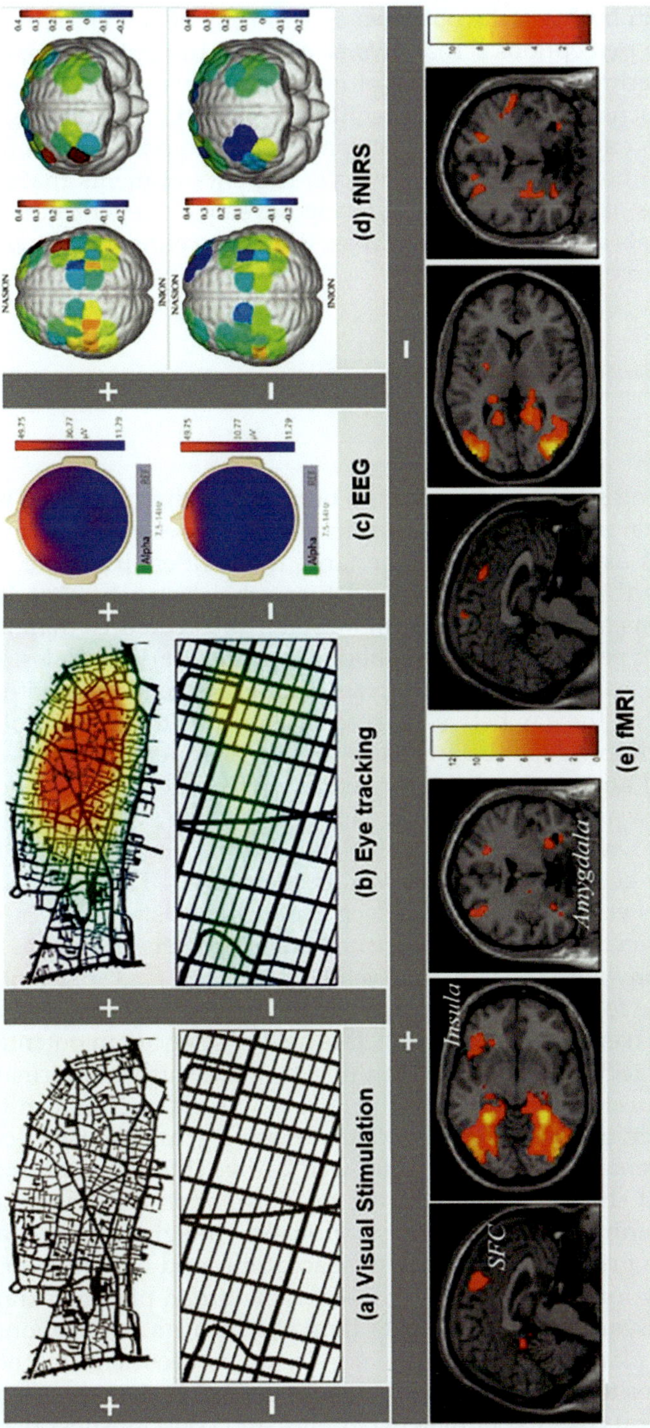

FIGURE 9.7
(Color online) Multimodal assessment of neurological responses to structural beauty.

Note: (a) Two urban grid structures with high (+) and low (–) livingness, (b) Eye-tracking heat maps showing gaze patterns, (c) EEG: Brainwave analysis highlighting emotional and cognitive responses, (d) fNIRS: Neural activation maps from visual stimuli, and (e) fMRI: Greater activation for complex (+) forms.

positive correlation between the objective assessments of spatial livingness and the subjective perceptions across diverse cultural and political contexts. Such studies underline the importance of integrating living structures into urban design, which highlights the potential of this approach to transform modern cities into engaging and vibrant spaces that promote human flourishing. Urban planners and architects who create environments that align with the principles of living structure can help develop a deeper connection between individuals and their surroundings, which can enhance the quality of life in urban areas.

9.6 Conclusion

The science of living structure represents a burgeoning discipline that transcends conventional mechanical frameworks and embraces an organic view of space that helps us understand and enhance the built environment. This approach stresses the intrinsic order and hierarchical structure underlying all spatial forms, marked by many more small substructures than large ones and repeating patterns of similar substructures on various scales. This perspective leverages principles of adaptation and differentiation with the intention of bridging the gap between art science and art to create a holistic framework that acknowledges beauty as an objective property that is rooted in the hierarchical and mathematical organization of space. This science is built on Tobler's law and the scaling law, which collectively posit that a living structure needs to have many more small elements than large ones, and that the elements must generally be similar within their levels. This approach is organic and recursive and exposes the deep order that exists in spaces, patterns and phenomena that exist in nature and cultivating environments that are in close harmony with the human psyche.

By unlocking the foundational principles of living structure and its applications, this paper has demonstrated how this field of science can potentially revolutionize architecture and urban planning, as well as urban informatics. Examining structural beauty through the lens of living structure can help transform urban environments in a way that fosters a greater sense of community, environmental sustainability, and well-being. The case studies that we have presented show the ability of the recursive approach to measure, differentiate and enhance the livingness of spaces. Integrating geospatial big data gives us a greater appreciation of space, time, and human mobility and paves the way for a comprehensive sustainable urban planning framework. Ongoing research will investigate living structure's implications in sustainable urban planning, geospatial analysis, and neuroscientific studies of human responses to spatial configurations, among other domains. By recognizing the integral role that human experience and consciousness play in the built environment, the science of living structure opens the door for an increasingly integrated and enriched approach to designing spaces that will promote human well-being and a place identity. This emerging field has the potential to unify artistic and scientific principles, ultimately creating environments that are not only functional, but also deeply fulfilling and inherently beautiful.

Acknowledgments

The chapter is a reprint of the paper by the same authors under the same title, originally published in Urban Informatics, with the kind permission of the publisher.

This chapter was prepared with the assistance of ChatGPT-4; however, we as the authors take full responsibility for any errors or oversights. The paper was supported by the Startup Fund of The Hong Kong University of Science and Technology (Guangzhou) (G0101000142), the City-University Joint Fund of the Science and Technology Project of Guangzhou (P00285), Guangdong Provincial Key Lab of Integrated Communication, Sensing and Internet of Things (2023B1212010007), the AI Research and Learning Base of Urban Culture under Project 2023WZJD008, and the Key Project of Jiangxi Provincial Natural Science Foundation (20242BAB26018). The chapter is a reprint of the paper by the same authors under the same title, originally published in Urban Informatics, with the kind permission of the publisher.

References

Alexander C. (2002–2005), *The Nature of Order: An Essay on the Art of Building and the Nature of the Universe*, Berkeley, CA: Center for Environmental Structure.

Alexander C., Ishikawa S., and Silverstein M. (1977), *A Pattern Language: Towns, Buildings, Construction*, New York: Oxford University Press.

Architectural Uprising (2014), *A People's Movement Against the Continued Uglification of Our Cities*, https://www.architecturaluprising.com/.

Architecture Education Declares (2019), *Open Letter to the Architectural Community: A Call for Curriculum Change*, http://livablecitylab.hkust-gz.edu.cn/ArchitectureEducationDeclares/

Bak P. (1996), *How Nature Works: The Science of Self-Organized Criticality*, New York: Springer-Verlag.

Bohm D. (1980), *Wholeness and the Implicate Order*, London and New York: Routledge.

Cuthbert A. R. (2007), Urban design: Requiem for an era – Review and critique of the last 50 years, *Urban Design International*, 12(4), 177–223.

Descartes R. (1644, 1983), *Principles of Philosophy*, Dordrecht: Reidel.

Ishizu T. and Zeki S. (2011), Toward a brain-based theory of beauty, *PLoS ONE*, 6(7), e21852.

Jacobs J. (1961), *The Death and Life of Great American Cities*, New York: Random House.

Jiang B. (2015a), Geospatial analysis requires a different way of thinking: The problem of spatial heterogeneity, *GeoJournal*, 80(1), 1–13.

Jiang B. (2015b), Wholeness as a hierarchical graph to capture the nature of space, *International Journal of Geographical Information Science*, 29(9), 1632–1648.

Jiang B. and de Rijke C. (2023), Living images: A recursive approach to computing the structural beauty of images or the livingness of space, *Annals of the Association of American Geographers*, 113(6), 1329–1347.

Jiang B. and Huang J. (2021), A new approach to detecting and designing living structure of urban environments, *Computers, Environment and Urban Systems*, 88, 101654.

Kawabata H. and Zeki S. (2004), Neural Correlates of Beauty, *Journal of Neurophysiology*, 91, 1699–1705.

Le Corbusier (1923), *Towards a New Architecture*, London: Butterworth Architecture.

Lewicka M. (2011), Place attachment: How far have we come in the last 40 years? *Journal of Environmental Psychology*, 31, 207–230.

Marshall S. (2012), Science, pseudo-science and urban design, *Urban Design International*, 17, 256–271.

Mehaffy M. W. and Salingaros N. A. (2020), The Chernobyl paradox: The intense connection between health and living structure, *Common Edge*, https://commonedge.org/the- chernobyl-paradox-the-intense-connection-between-health-and-living-structure/

Salingaros N. A. (1997), Life and complexity in architecture from a thermodynamic analogy, *Physics Essays*, 10, 165–173.

Salingaros N. A. and Sussman A. (2020), Biometric pilot-studies reveal the arrangement and shape of windows on a traditional façade to be implicitly "engaging", whereas contemporary façades are not, *Urban Science*, 4(2), 26, https://doi.org/10.3390/urbansci4020026

Seamon D. (2018), *Life Takes Place: Phenomenology, Life worlds, and Place Making*, London: Routledge.

Simon H. A. (1962), The architecture of complexity, *Proceedings of the American Philosophical Society*, 106(6), 467–482.

Sussman A. (2014), *Cognitive Architecture: Designing for How We Respond to the Built Environment*, London: Routledge.

Tobler W. (1970), A computer movie simulating urban growth in the Detroit region, *Economic Geography*, 46(2), 234–240.

Tuan Y. F. (1977), *Space and Place: The Perspective of Experience*, Minneapolis, MN: University of Minnesota Press.

Whitehead A. N. (1929), *Process and Reality: An Essay in Cosmology*, New York: Free Press.

10

Localised Sensor-Enhanced Assessment of Indoor Heat Exposure in Low-Income Housing in Global South

Ronita Bardhan, Jiayu Pan, and Maoran Sun

10.1 Introduction

Climate change triggers the variations in climatic conditions like temperature, precipitation, humidity, and wind, leading to more frequent extreme weather events such as heat waves or high-intensity summers. Despite the promise of Paris Agreement to limit global multiyear average warming to 1.5°C, in 2023, the annual mean surface temperature reached a record-breaking 1.45°C above pre-industrial levels, severely rising temperatures across the globe (Romanello et al., 2024). The extreme heat in many countries is reaching hazardous limits that can cause adverse health consequences. Although, it is evident that built environment is the largest modifier of the lethal heat, yet, what entails built environment interventions to mitigate or adapt to the extreme heat, remains understudied. Most built–environment-related heat stress studies focus on outdoor overheating and urban heat island (UHI) effects (Santamouris, 2020), utilising the land surface temperature retrieved from the satellite image and meteorological data from local weather station (Tham et al., 2020). Yet, warmer neighbourhoods were found to be associated with the higher deprivation level and severe health impacts (Hoffman et al., 2020; Santamouris, 2020; Voelkel et al., 2018). Recent studies indicate that increase of indoor temperature affects the heat-related mortality more than the increase of outdoor temperature (Santamouris, 2020) Hence, thermal environments of domestic spaces where people tend to spend the significant time needs a deep investigation.

Reliance on outdoor temperature to understand overheating could result in bias as the indoor temperature is not uniform due to the building characteristics and the access to mechanical cooling devices (Williams et al., 2019). One such domestic space where anomaly in indoor temperatures can elevate health risks in vulnerable population are social housing. Achieving a

DOI: 10.1201/9781003484219-10

stable indoor environment that does not threaten human survival can ensure reduced risk to health.

Social housing stock is characterised by high-density development, poor indoor environment quality, and thermal performance (Bardhan and Pan, 2023; Bardhan et al., 2024; Diaz Lozano Patino and Siegel, 2018; Haddad et al., 2022; Sunikka-Blank et al., 2021). Understanding indoor environmental condition requires data at fine-grained resolution, usually at unit and building levels, where local sensing can be included to understand the human experience, perception, intensity and duration of heat, and its subsequent consequences.

While the studies on the overheated environment and the impacts vary significantly across different study scales, and are mostly done in Europe and North America (McMichael et al., 2006), less attention has been paid to the indoor heat stress in Global South, especially in the low-income population who are more vulnerable to extreme temperature with less capacity to modify the environment and limited access to cooling resources (Voelkel et al., 2018).

This chapter explores the effects of indoor heat stress on low-income housing in the Global South, using advanced sensing techniques and data-driven methods to gain actionable insights. Through literature review and two case studies located in India and Ethiopia, the chapter seeks to highlight the critical need for sustainable, affordable strategies, and design improvements that can alleviate the burdens of heat stress on these communities. This study contributes to a growing body of research focused on urban resilience, emphasizing the importance of localized data to inform policy and design solutions tailored to the unique challenges of low-income neighbourhoods in the Global South.

Section 10.2 introduces the relevant literature to indoor overheating, identifying the potential consequences from the perspectives of health and energy. Section 10.3 presents a case of indoor heat in Ethiopian condominium housing schemes with simulation and interview evidence. Section 10.4 illustrates a case of exploring how urban heat waves is affecting household energy usage with local sensors in India.

10.2 The Impacts of Rising Indoor Temperature

Studies have identified the significant consequences of high outdoor and indoor temperature, such as the direct impact on air quality, human health and productivity, significant increase in the demand for cooling energy and rise of peak electricity demand, increased heat-related mortality, and intensified level of fatigue and mental health issue (Haddad et al., 2024; McMichael et al., 2006; Santamouris, 2020; Williams et al., 2012; Yenneti et al., 2020).

10.2.1 Heat and Health

The interrelationship between heat exposure and human body has been widely studied. The rising indoor temperature could impose significant threats on human health, by disturbing the human thermoregulation. Studies have shown that deviations from the optimal temperature range can lead to discomfort, reduced productivity, and general dissatisfaction with the indoor environment (Felgueiras et al., 2023; Yenneti et al., 2020). Long-term exposure to extreme temperature could cause cardiovascular, respiratory, and thermoregulation-related issues like stroke and exhaustion (Lima et al., 2020; Schwartz et al., 2004; Semenza, 1999), as well as cognitive and mental disorder (Mullins and White, 2019; Tham et al., 2020; Thompson et al., 2018). The negative effects of high indoor temperature are more commonly studied in the office literature. Working in a warm environment would lead to the feeling of mental fatigue and require more cerebral blood flow to maintain performance (Haddad et al., 2024; Tanabe et al., 2007).

The studies of urban overheating demonstrates that social demographic characteristics and urban landscape are closely linked to the mortality and morbidity risk (McMichael et al., 2006; Pyrgou and Santamouris, 2020; Uejio et al., 2011), especially during heat waves (Xu et al., 2013). High ambient temperature and high vulnerability subsequently links to a high mortality rate (Santamouris, 2020; Schinasi et al., 2018). The particularly vulnerable groups include the aging population, people with pre-existing health conditions, and those lack of access to financial and public support (Chow et al., 2012; Gosling et al., 2009; Peng Bi et al., 2011; Santamouris, 2020).

While housing is a key determinant of health (The Lancet, 2024), the extreme indoor condition is also examined in different housing settings. The studies in various countries report the observation of high indoor temperature in low-income dwellings during summer (Haddad et al., 2019; Krüger, 2015; Naicker et al., 2017; Williams et al., 2019). For example, in South Africa, government-built low-cost housing and informal settlement homes exhibited the greatest temperature gaps, with indoor temperatures being 4 to 5°C higher than the outdoor temperatures (Naicker et al., 2017). The need for improving housing quality and energy efficiency is emphasised to reduce the negative impact of inadequate indoor temperature (Lima et al., 2020; Santamouris and Kolokotsa, 2015).

10.2.2 Energy

The relationship between urban heatwaves and domestic energy usage has been the focus of significant scholarly attention (Haddad et al., 2024; Hatvani-Kovacs et al., 2016; Pyrgou and Santamouris, 2020). Research shows that urban heat waves lead to substantial increases in residential energy demand, primarily driven by the more frequent use of air conditioning and other cooling mechanisms. For instance, studies conducted in global cities such as Los Angeles, Tokyo, and Bangkok demonstrate that energy consumption can surge by 8.5% per degree of temperature rise during heatwave periods as residents seek to maintain indoor comfort levels (Santamouris and Kolokotsa, 2015). This demand spike places stress on urban power grids, leading to challenges in maintaining energy reliability and affordability. The phenomenon is particularly pronounced in dense urban areas, where the UHI effect further amplifies ambient temperatures, thereby compounding the energy needs for cooling (Akbari et al., 2001; Oke, 1982).

Moreover, socioeconomic disparities play a critical role in mediating the effects of urban heatwaves on energy usage. Households in lower-income communities often lack access to efficient cooling technologies or are deterred by the cost of electricity, leading to a reliance on passive measures or an inability to mitigate heat stress effectively (Sun et al., 2021). Such disparities have important implications for public health and energy equity, as vulnerable populations are more likely to suffer from heat-related illnesses during extreme temperature events. As cities pursue redevelopment and energy-efficient retrofits, incorporating resilient design features that address heat stress is crucial to reducing energy burdens and enhancing thermal comfort in residential spaces (Shen et al., 2024). This body of literature underscores the need for integrative approaches that combine urban planning, energy management, and social policy to address the dual challenges of urban heatwaves and sustainable energy use.

While the relationship between heatwave and household energy usage has been studied extensively, less attention has been paid to the indoor heat stress in Global South, especially towards the low-income population who are more vulnerable to extreme temperature with less capacity to modify the environment and limited access to cooling resources.

10.3 Case Study 1: Social Sensing of Indoor Heat in IHDP Housing, Ethiopia

10.3.1 Background

With the rapid urbanisation in Ethiopia at a national rate of 4% to 6% per year (Gardner, 2017), affordable housing schemes are introduced to accommodate the increasing urban population. Integrated Housing and Development Plan (IHDP), supported by Ethiopian government, is launched to supply affordable housing to low- and middle-income dwellers, aiming to build 400,000 housing units, creating 200,000 job opportunities and reducing the slum areas by 50% (UN-HABITAT, 2010; Workineh, 2022).

Figure 10.1 shows the massive IHDP housing developments in Addis Ababa and Jimma, Ethiopia. These new-built high-density high-rise concrete housing are often criticised with its failure to accommodate and sustain local people's traditional way of life, causing loneliness and socioeconomic segregation (Gardner, 2017). Many condominium housing sites are located

FIGURE 10.1
IHDP housing developments (condominiums) in Addis Ababa and Jimma, Ethiopia.

Source: Photos by the authors and Belsti Asres.

on the edge of the cities with limited access to employment opportunities (Sunikka-Blank et al., 2021; UN-HABITAT, 2010).

The IHDP sites studied in this section are located in Jimma City, Ethiopia. Jimma (7°40′N, 36°50′E) is the largest city in southwestern Oromia Region, at 352km away from the capital Addis Ababa (Dibaba, 2018). The city locates in a tropical monsoon climate zone (Köppen classification: Am), with a long rainy season from March to October.

10.3.2 Methods

The study investigates the extreme and chronic heat stress experienced by the IHDP occupants and explores the implications of building design and household practice norms in mitigating the excess heat stress, through qualitative and quantitative evidence collection and analysis. Simulation and interview narratives are combined to understand the implications of rising indoor temperature on occupants' living condition, experience, and adaptive measures.

10.3.2.1 Simulation

A simulation-based evaluation of the overheating intensity was conducted with the projected climate change scenarios in four years 2018, 2020, 2050, and 2080. The projection of future climate is generated by climate change world weather file generator (CCWorldWeatherGen) (Jentsch et al., 2013). The input EnergyPlus Weather (epw) file of Jimma City was downloaded from Weather Data, EnergyPlus (EnergyPlus, 2023). The A2 storyline and scenario was selected from climate change scenario data, as it describes a heterogeneous world with continuously increasing population, regionally-oriented economic growth, and fragmented technological change, and represents a 'medium-high' emission scenario (IPCC, 2000).

Two typical types of buildings were selected to run simulation. The characteristics of the buildings are shown in Table 10.1. They are both G+4 buildings

TABLE 10.1

Building Features in Simulation

	Type A	Type B
Number of units	25	25
Floor height	10 feet (approximately 3 metre)	
Wall	Hollow concrete blocks (HCB), plastered and painted	
Roof	EGA 4000 roof cover, 4mm thick galvanised iron sheet	
Floor	Cement screed	
Window	Aluminium or metal framed, 4mm single glazed non-insulated glass Large window: 120 × 160 cm Small window: 80 or 90 × 60 cm	
Door	Half glass metal door Size: 90 × 200 cm	
Units (Top view)	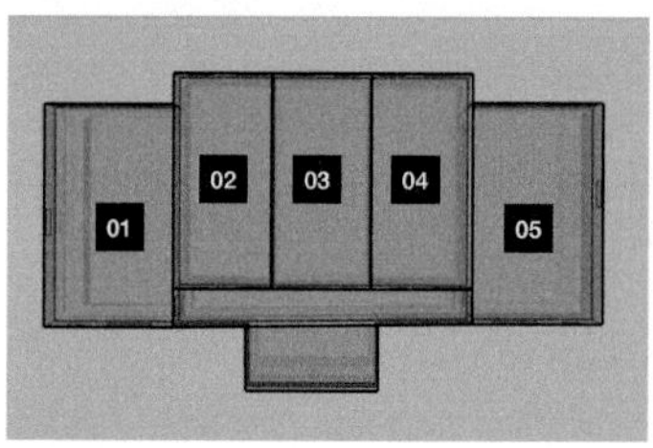	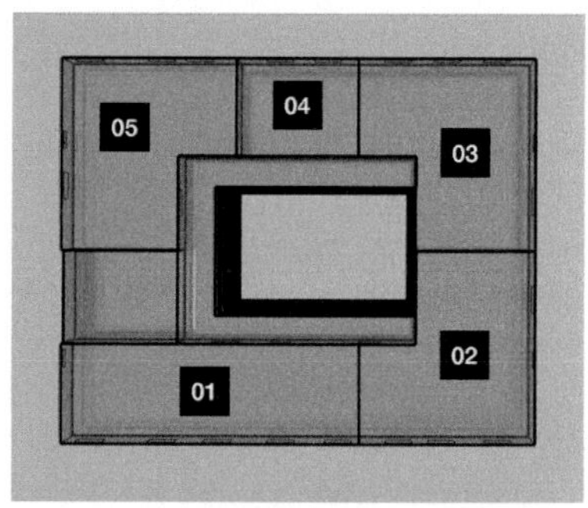
3-dimension model	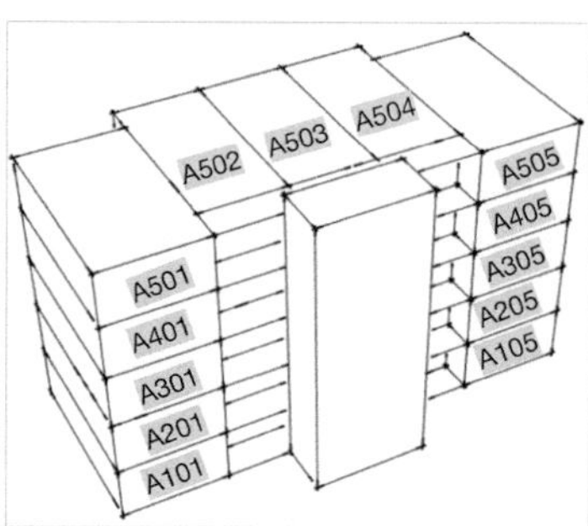	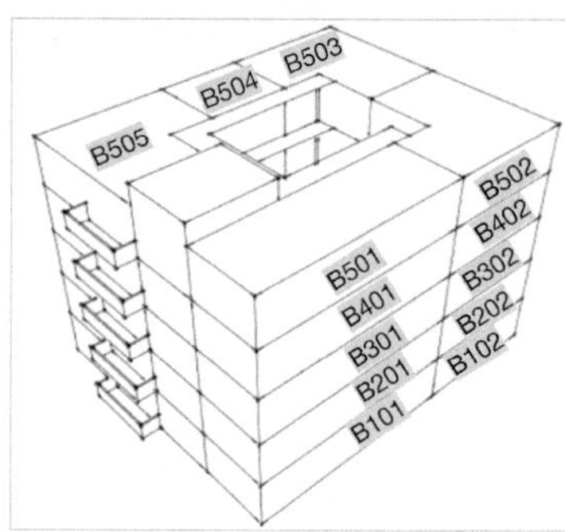

with 25 units. Type A has a linear corridor with a staircase attached in the middle of the corridor, while type B has a circular corridor with an inner courtyard. The base models of the selected buildings are built in SketchUp and the simulations are run via OpenStudio plug-in (Guglielmetti et al., 2011). OpenStudio is an integrated open-source platform that supports the building energy modelling using EnergyPlus. The primary outputs from the simulation include indoor temperatures and heat index (HI) hours. HI is a measure that reflects the expected stress on the human body due to the combined effects of humidity and temperature (Steadman, 1979). It is categorised into six levels: safe (≤ 26.7°C), caution (26.7 to 32.2°C), extreme caution (32.2–39.4°C), danger (39.4–51.7°C), and extreme danger (>51.7°C). The physiological responses to these categories vary, ranging from slight fatigue under caution conditions to a high risk of heat stroke in extreme danger conditions.

10.3.2.2 Interview

Qualitative-based videography interviews have been conducted to capture the experiences and perceptions of occupants living in slum rehabilitation housing, with the theme of 'What does "heat" mean to the residents in IHDP?'. These narratives provide valuable insights into household behaviours, practices, and norms regarding heat adaptation and mitigation.

A total of 25 households were interviewed in mid-August 2022. During the interview process, the indoor and outdoor temperatures are measured, with the range of 25–27°C and 17–19°C, respectively. The households are sampled across three IHDP sites (Kochi, Yetebaberut, and Kito) around Jimma, as shown in Table 10.2. To ensure the diversity of interviews, these households spread across various types of units, ranging from studio and one-bedroom to three-bedrooms flats, and across different floors. The interview involves 14 female and 11 male participants. Most of the males are government employee or private worker, while most of the females are housewife. Their average family size is 4, and the majority of households have 4 to 5 members.

The interview is semi-structured, guided by a set of open-ended questions:

How many hours do you typically spend indoors each day?

How do you perceive the heat in your home? At what point does it become uncomfortable or burdensome?

What strategies do you use to cope with the heat in your home?

During which season or time of day do you find the indoor temperature to be most uncomfortably warm?

How does your experience with heat compare before and after moving into the condominium?

In what ways do you think the building materials used in your home contribute to the heat inside?

TABLE 10.2

IHDP Sites Selected for Interview

	Yetebaberut—EDDC Condo Site	Kito Condo Site	Kochi Residency
Number of units around the site	176	504	97
Number of interviewed households	5	15	5
Number of floors	G+2 floors	G+4 floors	G+2 floors
Site photo			

Do you think it will get hotter in the coming years? What solutions or technologies, if any, do you have to manage the heat?

How effective are the measures you take to reduce the negative impact of heat in your home?

The interviews were originally conducted in Amharic, which is the most widely spoken language and the official language of the country. Later the transcripts were translated to English by researchers for analysis.

10.3.3 Results and Analysis

10.3.3.1 Quantitative Analysis

Figure 10.2 shows the variations of indoor and outdoor temperatures across the year with the projection in 2020, 2050, and 2080. Figure 10.3 illustrates a comparison of average indoor temperatures between two simulated building types. The data reveals a substantial difference between outdoor and indoor temperatures, with an average discrepancy of 10°C. This suggests that indoor temperatures may often remain within unsafe ranges throughout the year, and the outdoor temperature would not be a sufficient indicator to assess the overheating condition experienced in the condominium housing. Projections indicate a significant rise in outdoor temperatures, from an annual average of

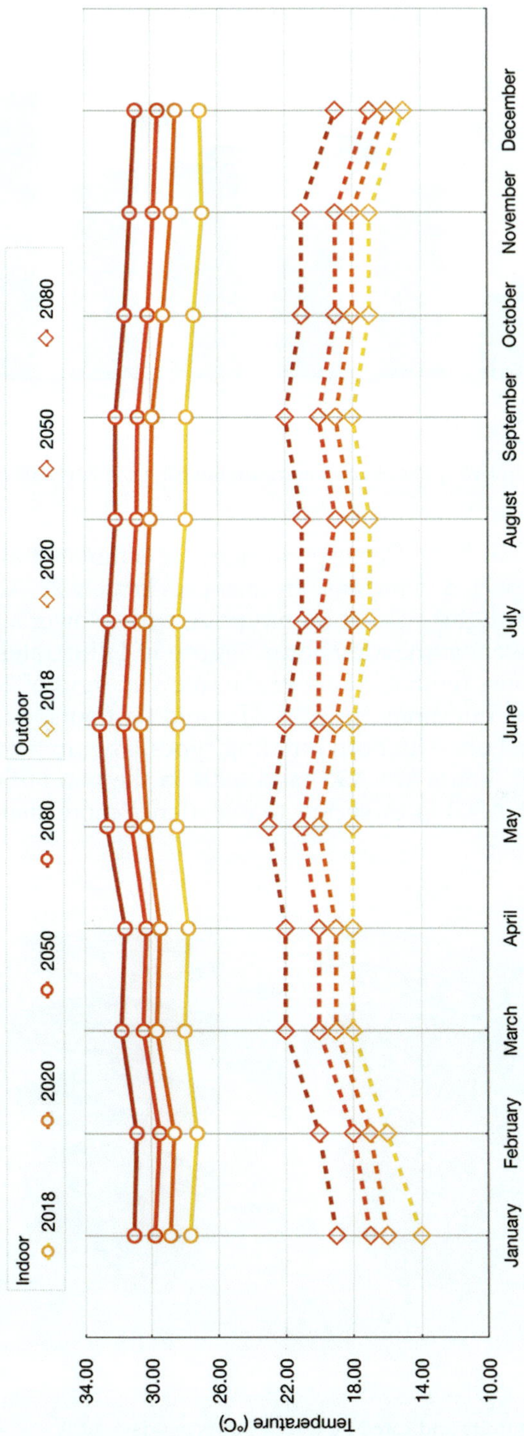

FIGURE 10.2
Monthly variations of indoor and outdoor temperatures in Jimma, Ethiopia.

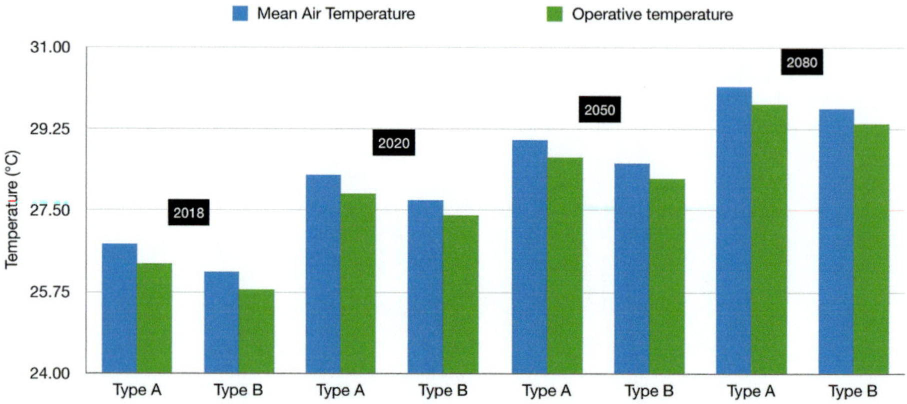

FIGURE 10.3
Temperature variations in two types of condominium buildings, in 2018, 2020, 2050, and 2080.

17°C in 2018 to 21°C in 2080. Correspondingly, the simulated average indoor temperatures are expected to increase from approximately 28–32°C.

Building type B is slightly cooler than type A, with a lower average indoor temperature and fewer overheating hours. Figure 10.4 illustrates the distribution of unsafe HI hours (with a HI above the safety margin of 26.7°C) among the units in the two buildings in 2020. The middle floors, particularly the second floor, are the hottest in both building types. The units with the highest 'Extreme Danger' hours are A303 and B404 in the two buildings respectively. By 2080, unit A303 is projected to experience more than 21 hours of

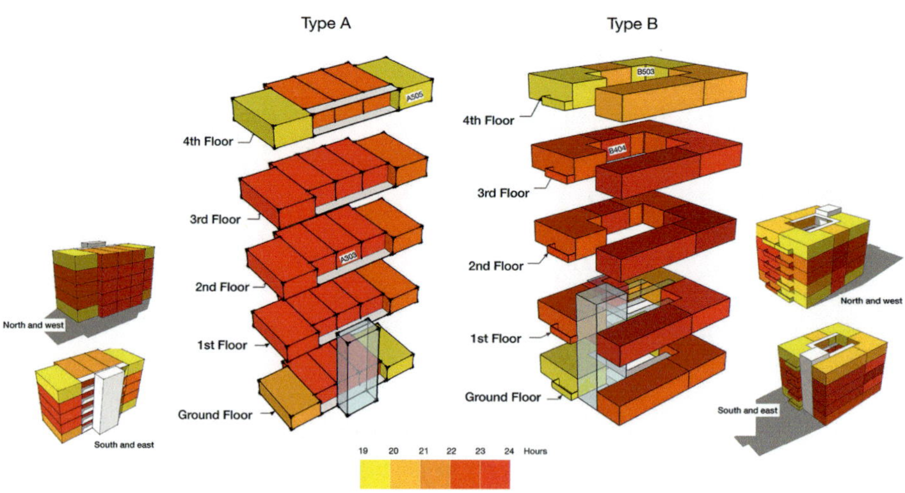

FIGURE 10.4
Heat vulnerability of the units (indicated by unsafe hours per day), 2020.

extreme danger conditions per day, while unit B404 will face 13.6 hours. Conversely, the units with the most hours in the safe range are A505 and B503.

10.3.3.2 Qualitative Narratives

The exploration of qualitative narratives from interviews provides a human-centric perspective that is essential for informing heat-resilient affordable housing design. The narrative analysis reveals a complex picture of how residents in IHDP condominium experience and cope with indoor heat stress. Their stories provide valuable insights into the everyday realities of living in these conditions and point to specific areas where intervention and improvements are needed.

Many residents report feeling warm or hot, particularly during the nighttime and when staying indoors during autumn. This perception often leads to discomfort, sweating, fatigue, and stress. Difficulty sleeping due to heat is a recurring issue; over half of the 25 interviewees directly mentioned that indoor heat affects their sleep at night.

"Indoor heat is seasonal, but mostly I feel warm at nighttime, and it affects my sleep. The heat becomes a burden in autumn, especially in the afternoon and nighttime." (Interview 23)

The coping strategies reported by interviewees are basic, including opening doors and windows to increase airflow, sitting in corridors, and adjusting clothing. However, these measures are often impractical due to external factors such as mosquitoes, rats, or concerns for children's safety. Some residents also mentioned taking showers, particularly before sleeping, as a method to cool down. Most respondents indicated that they do not have access to mechanical cooling devices. Only one interviewee mentioned considering mechanical cooling if the heat becomes unbearable. Female, especially a working female, adjusts her cooking time to morning:

"Sometimes I face sleeping problems and stress at night. I open the window and cook in the morning to cope with it. But at night, I can't open windows because of mosquitoes, so I may take a shower instead." (Interview 6)

When comparing their experiences before and after moving to the condominium housing, 19 interviewees noted that the condominium housing is warmer. In Ethiopia, typical low-income dwellings use wattle and daub ('chikka') walls, with roof rafters made of round tree lengths (usually Eucalyptus), covered with corrugated iron sheeting, and floors made of skim concrete or compacted earth. In contrast, the condominium blocks are constructed with reinforced concrete frames and slab construction, with hollow or fired brick infill walls. The modular design, aimed at reducing construction costs and time, uses standard materials and includes metal-framed, single-glazed windows and doors (UN-HABITAT, 2010).

The narratives highlight how the building environment, including factors like orientation, sun exposure, materials, and lack of open space, significantly influences residents' experiences with heat. Many residents believe that the materials and design of the condominium housing contribute to indoor heat, particularly due to poor-quality roofing, inadequate ventilation, and a lack of external shading. Direct sun exposure is frequently mentioned as exacerbating indoor heat, especially for upper-level units.

"I strongly believe the building construction materials trap heat. The house is designed only for accessibility; it also has orientation problems." (Interview 21)

"This block is closed and does not allow natural ventilation. Temperature-sensitive materials, like concrete block walls, trap heat. There is also a problem with building orientation regarding wind direction, and even the upper windows are locked." (Interview 11)

Among all interviews, only one person who lives on the ground floor reported no significant issue with heat—*"This house is located on the ground level with no direct sun exposure, and I think that helps keep it cool."* (Interview 3)

Many residents also express concerns that indoor heat will increase in the future, and there is limited optimism about the effectiveness of current coping mechanisms or the availability of new solutions. A few interviewees suggest adding more green spaces as a potential solution.

The perspectives and challenges of female residents are particularly notable as many women participate in the interview. Ethiopian women often occupy marginalised positions with lower education levels and less participation in professional work compared to men. Many housewives in the interview reported spending over 16 hours indoors, with some spending the entire day inside their units. They highlighted concerns about cooking activities, children's safety, and inappropriate kitchen arrangements. These narratives underscore that indoor overheating disproportionately affects the health and well-being of women, indicating a need for further studies focused on their specific experiences with heat stress (Bardhan et al., 2023; Van Daalen et al., 2024).

10.4 Case Study 2: Local Sensing of Energy Usage and Heat Stress in Mumbai, India

10.4.1 Background

Over 23% of the world's population resides in informal settlements (United Nations, 2019). Informal settlements represent an unsustainable form of habitation due to a lack of basic services like sanitation, water, electricity, and healthcare, and are characterised by overcrowded and unsafe building structures. These conditions increase residents' vulnerability to urban heatwaves, which are becoming more frequent due to climate change. As a result, there has been a massive worldwide push for the redevelopment of slum communities. For example, as part of their 'Housing for All' programme, the central Indian government is advancing redevelopment plans for Dharavi, Mumbai—Asia's largest informal settlement.

Dharavi's redevelopment will have substantial impacts on the energy sustainability of the community and the city of Mumbai, as its over 1 million residents contribute an estimated $1 billion/year to the city's GDP (Jana et al., 2016). Redevelopment design decisions currently being made with respect to spatial morphologies, community layouts, and building materials will have significant long-term implications for energy demand, the integration of renewable energy, and resilience to urban heat waves over the next several decades.

Urban heat waves present a dual challenge by increasing indoor temperatures and elevating the risk of heat stress among residents, particularly in densely populated areas like Dharavi. Currently, redevelopment plans are progressing without comprehensive analysis and assessment of their implications on overall energy efficiency, renewable energy integration, and, most critically, the burdens to be borne by Dharavi residents.

To address the challenge of understanding when and how Dharavi residents consume energy, as well as the impact of heat stress on their energy usage patterns, this project deployed a series of smart meter sensors across selected households in Dharavi, Mumbai (Figure 10.5). These smart meters record real-time energy consumption, providing detailed insights into how residents respond to heatwaves and manage energy demands under varying conditions.

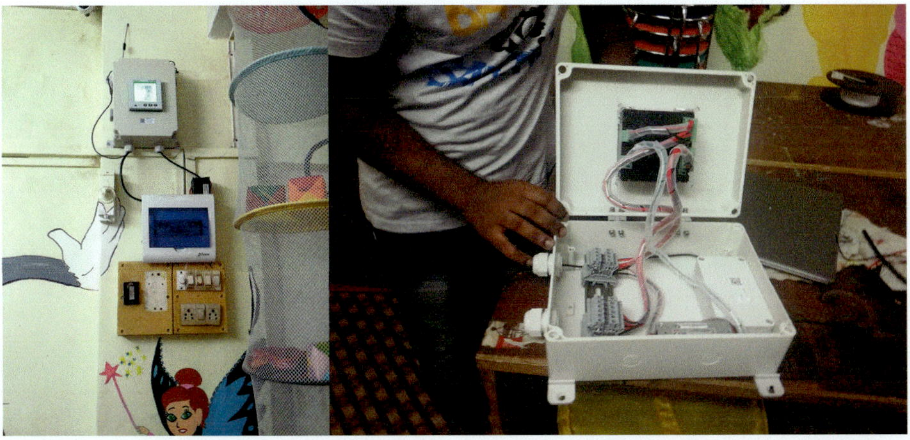

FIGURE 10.5
Installed smart meter sensors.

10.4.2 Data

10.4.2.1 Weather Data

Hourly weather station data was utilised to provide comprehensive environmental context for energy consumption patterns within households in Dharavi. The weather station recorded key meteorological variables including temperature, humidity, feels-like temperature, dew point temperature, and a description of weather condition. These measurements were collected at hourly intervals to capture diurnal variations and to understand their relationship with residential energy demand.

Weather data was obtained from the Visual Crossing data platform (Visual Crossing, 2023) using the weather station 'BOMBAY COLABA, IN' located at coordinates 18.9°N, 72.817°E. To align with the timeframe of the smart meter data, weather data covering the period from August 1, 2019, to March 31, 2024, was downloaded. Figure 10.6 presents the monthly average temperature over the project period, providing a contextual baseline for analysing heat stress impacts on energy consumption.

10.4.2.2 Smart Meter Data

To address the gap in energy consumption data within informal settlements, we deployed a series of smart meter sensors across selected households in Dharavi, Mumbai. These sensors capture granular data on electricity usage, providing insight into both temporal and spatial consumption patterns within residential units. The smart meters are configured to record energy use at 1-minute intervals, enabling the detailed analysis of peak demand

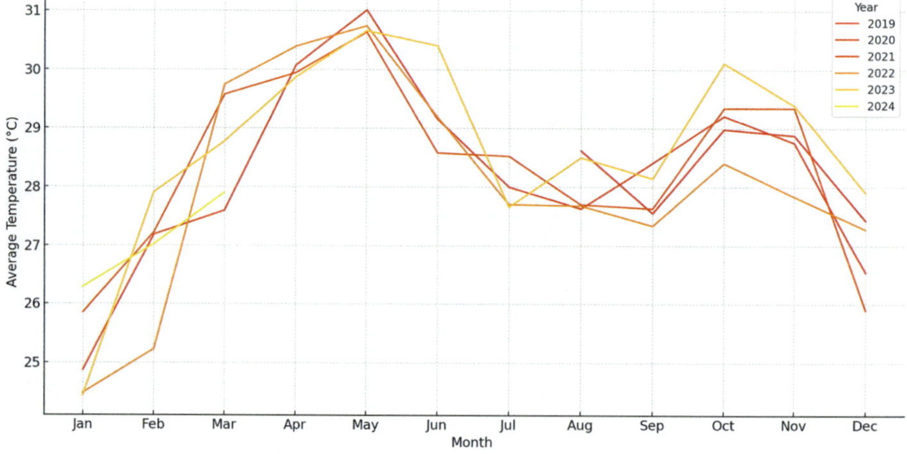

FIGURE 10.6
Monthly variations in temperature across different years.

TABLE 10.3

Data Collection Periods for Each Household

Household	Start Date	End Date
House 1	08/22/2019	09/11/2020
House 2	08/23/2019	09/11/2020
House 3	03/24/2020	03/27/2024
House 4	03/24/2020	02/05/2022
House 5	03/24/2020	03/25/2024

periods, variability in energy use, and the impacts of external factors such as urban heatwaves on indoor energy behaviour.

The collected data encompasses aggregate consumption metrics, detailed load profiles across various times of day, and electrical parameters including R, Y, and B phase voltages, as well as average voltage levels. Additional metrics captured include R, Y, B phase and neutral currents, active current, and forward active energy. Due to constraints imposed by the COVID-19 pandemic, smart meters were installed across households at different times. Table 10.3 provides an overview of the specific data collection periods for each household.

10.4.3 Methods

10.4.3.1 Heat Stress Calculation

To quantify the impact of heat on household energy consumption, the HSI is calculated based on hourly weather station data, which includes ambient temperature and relative humidity. The HSI provides an estimate of the physiological stress experienced by individuals under different environmental conditions and is commonly used to assess the potential for heat-related health risks.

We employed the following formula to calculate the HSI:

$$HSI = T - 0.55 \times (1 - RH) \times (T - 14.5)$$

where T represents the ambient temperature in degrees Celsius, and RH is the relative humidity expressed as a decimal. This calculation captures the combined effect of temperature and humidity on perceived heat, which can trigger increased energy use as residents respond with cooling measures.

10.4.3.2 Regression Analysis

To explore the relationship between household energy consumption and environmental conditions, a regression analysis was conducted using the combined dataset of smart meter readings and calculated HSI. The analysis aimed to quantify the impact of external meteorological factors on residential energy use, with a particular focus on heat stress as a key predictor.

The dependent variable in the regression model was hourly energy consumption (in kilowatt-hours), while independent variables included hourly temperature, relative humidity, and interaction terms to account for potential synergistic effects. All data were pre-processed to ensure temporal alignment, with missing values imputed using a time-series-based approach.

10.4.4 Results

10.4.4.1 Heat Stress Calculation

The analysis of HSI data reveals that approximately 43% of recorded HSI values exceed 27 (Figure 10.7), indicating moderate to high heat stress levels for about half of the observation period. An HSI value above 27 is typically associated with increased discomfort and potential health risks, suggesting that a substantial portion of time was characterized by conditions likely to impact resident well-being and prompt greater energy use for cooling.

10.4.4.2 Regression Analysis

The regression analysis aimed to investigate the impact of temperature and humidity on household energy consumption. Results indicate a positive association between temperature and energy consumption, with a statistically significant regression ($p < 0.001$). This suggests that for an increase in

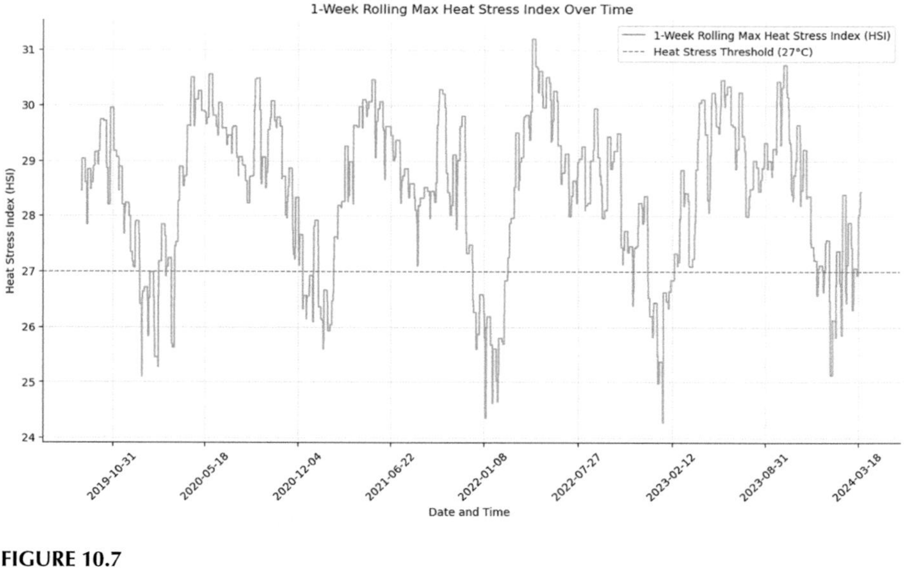

FIGURE 10.7
Week rolling max calculated heat stress index over time.

temperature, energy consumption also increases, likely due to the increased use of cooling devices during higher temperatures. Humidity also exhibited a significant positive relationship with energy consumption ($p < 0.001$), indicating that higher humidity levels lead to a moderate increase in energy demand.

Using the calculated HSI values, we isolated the regression results for heatwave periods. The analysis demonstrated a significant positive correlation between temperature and energy consumption during these periods, underscoring the heightened demand for active cooling in response to elevated temperatures. This finding highlights the increased energy burden on households during heatwaves, as rising temperatures drive greater reliance on cooling systems.

10.5 Conclusion

This chapter demonstrates the local sensing of heat with the case studies in social housing in Global South. It is crucial to understand the implication of heat not just at the large-scale but at a more local level with human-centric data, to develop corresponding strategies to support the adaptation and mitigation of overheating issue in the face of climate change. The local sensing of heat, as introduced in this chapter, directly align with several Sustainable Development Goals (SDGs) (United Nations, 2023), including Goal 3 (Good health and well-being), Goal 7 (Affordable and clean nergy), and Goal 11 (Sustainable cities and communities).

The quantitative and qualitative evaluation of the overheating condition in both Ethiopian and Indian affordable housing clearly indicate the importance of addressing indoor heat stress in the massive social housing schemes constructed in Global South, while the building design could play a key role in it (Bardhan, 2024). From the local-sensing evidence, the need for improving the design of condominium housing to protect residents from indoor heat stress is highlighted, which further calls for the need for planning and government agencies to play a role in bridging the gaps between climate change and rapid urbanisation in less developed countries (Mohammed et al., 2019).

Acknowledgement

Part of the work presented in this chapter is supported by the research projects 'HEROES - Health and Energy impacts of Residential Overheating in Ethiopian Slum rehabilitation housing', funded by Cambridge Africa Alborada Research Fund and 'Energy use in Ethiopian low-income housing' funded by Energy Interdisciplinary Research Centre, University of Cambridge. JP and

RB would like to acknowledge Prof. Esayas Alemayehu and Belsti Asres from Jimma University for their contributions in the data collection. JP would also like to thank St John's College, University of Cambridge for supporting the fieldwork in Ethiopia with Wilberforce and Clarkson (Williams) Travel Fund. RB acknowledges the support Faclon Labs (https://www.faclon.com/) for enabling data collection in Mumbai. RB extends thanks to Ms Krithika Panicker for conducting the field work supervision in Mumbai.

References

Akbari, H., Pomerantz, M., Taha, H., 2001. Cool surfaces and shade trees to reduce energy use and improve air quality in urban areas. Sol. Energy 70, 295–310. https://doi.org/10.1016/S0038-092X(00)00089-X

Bardhan, R., 2024. Climate adaptation: How can design science help the transition? https://doi.org/10.33774/coe-2024-hvnmn

Bardhan, R., Debnath, R., Mukherjee, B., 2023. Factor in gender to beat the heat in impoverished settlements. Nature 620, 727. https://doi.org/10.1038/d41586-023-02632-3

Bardhan, R., Pan, J., 2023. The Why? How? What? and What-ifs of Mass Slum Rehabilitation Housing in India, in: Informal Settlements of the Global South. Routledge, London, pp. 254–275. https://doi.org/10.4324/9781003191407-18

Bardhan, R., Pan, J., Chen, S., Cho, T.Y., 2024. Breathing space in a compact city: Impacts of urban re-densification on Mumbai's low-income housing environment. Habitat Int. 149, 103098. https://doi.org/10.1016/j.habitatint.2024.103098

Chow, W.T.L., Chuang, W.-C., Gober, P., 2012. Vulnerability to extreme heat in metropolitan Phoenix: Spatial, temporal, and demographic dimensions. Prof. Geogr. 64, 286–302. https://doi.org/10.1080/00330124.2011.600225

Diaz Lozano Patino, E., Siegel, J.A., 2018. Indoor environmental quality in social housing: A literature review. Build. Environ. 131, 231–241. https://doi.org/10.1016/j.buildenv.2018.01.013

Dibaba, W.T., 2018. A review of sustainability of urban drainage system: traits and consequences/REVISÃO DA SUSTENTABILIDADE DE UM SISTEMA DE DRENAGEM URBANA: CARACTERÍSTICAS E CONSEQÜÊNCIAS. J. Sediment. Environ. 3, 131–137. https://doi.org/10.12957/jse.2018.37825

EnergyPlus, 2023. Weather Data [WWW Document]. https://energyplus.net/weather

Felgueiras, F., Mourão, Z., Moreira, A., Gabriel, M.F., 2023. Indoor environmental quality in offices and risk of health and productivity complaints at work: A literature review. J. Hazard. Mater. Adv. 10, 100314. https://doi.org/10.1016/j.hazadv.2023.100314

Gardner, T., 2017. "Addis has run out of space": Ethiopia's radical redesign. The Guardian. https://www.theguardian.com/cities/2017/dec/04/addis-ababa-ethiopia-redesign-housing-project

Gosling, S.N., Lowe, J.A., McGregor, G.R., Pelling, M., Malamud, B.D., 2009. Associations between elevated atmospheric temperature and human mortality: a critical review of the literature. Clim. Change 92, 299–341. https://doi.org/10.1007/s10584-008-9441-x

Guglielmetti, R., Macumber, D., Long, N., 2011. OpenStudio: An open source integrated analysis platform; preprint, in: Proceedings of Building Simulation 2011: 12th Conference of International Building Performance Simulation Association.

Haddad, S., Paolini, R., Synnefa, A., De Torres, L., Prasad, D., Santamouris, M., 2022. Integrated assessment of the extreme climatic conditions, thermal performance, vulnerability, and well-being in low-income housing in the subtropical climate of Australia. Energy Build. 272, 112349. https://doi.org/10.1016/j.enbuild.2022.112349

Haddad, S., Pignatta, G., Paolini, R., Synnefa, A., Santamouris, M., 2019. An extensive study on the relationship between energy use, indoor thermal comfort, and health in social housing: the case of the New South Wales, Australia. IOP Conf. Ser. Mater. Sci. Eng. 609, 042067. https://doi.org/10.1088/1757-899X/609/4/042067

Haddad, S., Zhang, W., Paolini, R., Gao, K., Altheeb, M., Al Mogirah, A., Bin Moammar, A., Hong, T., Khan, A., Cartalis, C., Polydoros, A., Santamouris, M., 2024. Quantifying the energy impact of heat mitigation technologies at the urban scale. Nat. Cities 1, 62–72. https://doi.org/10.1038/s44284-023-00005-5

Hatvani-Kovacs, G., Belusko, M., Pockett, J., Boland, J., 2016. Assessment of Heatwave Impacts. Procedia Eng. 169, 316–323. https://doi.org/10.1016/j.proeng.2016.10.039

Hoffman, J.S., Shandas, V., Pendleton, N., 2020. The effects of historical housing policies on resident exposure to intra-urban heat: A study of 108 US urban areas. Climate 8, 12. https://doi.org/10.3390/cli8010012

IPCC (Ed.), 2000. Emissions scenarios: Summary for policymakers; A special report of IPCC Working Group III$Intergovernmental Panel on Climate Change, IPCC special report. Intergovernmental Panel on Climate Change.

Jana, A., Bardhan, R., Sarkar, S., Kumar, V., 2016. Framework to assess and locate affordable and accessible housing for developing nations: Empirical evidences from Mumbai. Habitat Int. 57, 88–99. https://doi.org/10.1016/j.habitatint.2016.07.005

Jentsch, M.F., James, P.A.B., Bourikas, L., Bahaj, A.S., 2013. Transforming existing weather data for worldwide locations to enable energy and building performance simulation under future climates. Renew. Energy 55, 514–524. https://doi.org/10.1016/j.renene.2012.12.049

Krüger, E.L., 2015. Urban heat island and indoor comfort effects in social housing dwellings. Landsc. Urban Plan. 134, 147–156. https://doi.org/10.1016/j.landurbplan.2014.10.017

Lima, F., Ferreira, P., Leal, V., 2020. A review of the relation between household indoor temperature and health outcomes. Energies 13, 2881. https://doi.org/10.3390/en13112881

McMichael, A.J., Woodruff, R.E., Hales, S., 2006. Climate change and human health: Present and future risks. The Lancet 367, 859–869. https://doi.org/10.1016/S0140-6736(06)68079-3

Mohammed, M.U., Hassan, N.I., Badamasi, M.M., 2019. In search of missing links: Urbanisation and climate change in Kano Metropolis, Nigeria. Int. J. Urban Sustain. Dev. 11, 309–318. https://doi.org/10.1080/19463138.2019.1603154

Mullins, J.T., White, C., 2019. Temperature and mental health: Evidence from the spectrum of mental health outcomes. J. Health Econ. 68, 102240. https://doi.org/10.1016/j.jhealeco.2019.102240

Naicker, N., Teare, J., Balakrishna, Y., Wright, C., Mathee, A., 2017. Indoor temperatures in low cost housing in Johannesburg, South Africa. Int. J. Environ. Res. Public. Health 14, 1410. https://doi.org/10.3390/ijerph14111410

Oke, T.R., 1982. The energetic basis of the urban heat island. Q. J. R. Meteorol. Soc. 108, 1–24. https://doi.org/10.1002/qj.49710845502

Peng Bi, Williams, S., Loughnan, M., Lloyd, G., Hansen, A., Kjellstrom, T., Dear, K., Saniotis, A., 2011. The effects of extreme heat on human mortality and morbidity in Australia: Implications for public health. Asia Pac. J. Public Health 23, 27S–36S. https://doi.org/10.1177/1010539510391644

Pyrgou, A., Santamouris, M., 2020. Probability risk of heat- and cold-related mortality to temperature, gender, and age using GAM Regression Analysis. Climate 8, 40. https://doi.org/10.3390/cli8030040

Romanello, M., Walawender, M., Hsu, S.-C., Moskeland, A., Palmeiro-Silva, Y., Scamman, D., Ali, Z., Ameli, N., Angelova, D., Ayeb-Karlsson, S., Basart, S., Beagley, J., Beggs, P.J., Blanco-Villafuerte, L., Cai, W., Callaghan, M., Campbell-Lendrum, D., Chambers, J.D., Chicmana-Zapata, V., Chu, L., Cross, T.J., Van Daalen, K.R., Dalin, C., Dasandi, N., Dasgupta, S., Davies, M., Dubrow, R., Eckelman, M.J., Ford, J.D., Freyberg, C., Gasparyan, O., Gordon-Strachan, G., Grubb, M., Gunther, S.H., Hamilton, I., Hang, Y., Hänninen, R., Hartinger, S., He, K., Heidecke, J., Hess, J.J., Jamart, L., Jankin, S., Jatkar, H., Jay, O., Kelman, I., Kennard, H., Kiesewetter, G., Kinney, P., Kniveton, D., Kouznetsov, R., Lampard, P., Lee, J.K.W., Lemke, B., Li, B., Liu, Y., Liu, Z., Llabrés-Brustenga, A., Lott, M., Lowe, R., Martinez-Urtaza, J., Maslin, M., McAllister, L., McMichael, C., Mi, Z., Milner, J., Minor, K., Minx, J., Mohajeri, N., Momen, N.C., Moradi-Lakeh, M., Morrisey, K., Munzert, S., Murray, K.A., Obradovich, N., O'Hare, M.B., Oliveira, C., Oreszczyn, T., Otto, M., Owfi, F., Pearman, O.L., Pega, F., Perishing, A.J., Pinho-Gomes, A.-C., Ponmattam, J., Rabbaniha, M., Rickman, J., Robinson, E., Rocklöv, J., Rojas-Rueda, D., Salas, R.N., Semenza, J.C., Sherman, J.D., Shumake-Guillemot, J., Singh, P., Sjödin, H., Slater, J., Sofiev, M., Sorensen, C., Springmann, M., Stalhandske, Z., Stowell, J.D., Tabatabaei, M., Taylor, J., Tong, D., Tonne, C., Treskova, M., Trinanes, J.A., Uppstu, A., Wagner, F., Warnecke, L., Whitcombe, H., Xian, P., Zavaleta-Cortijo, C., Zhang, C., Zhang, R., Zhang, S., Zhang, Y., Zhu, Q., Gong, P., Montgomery, H., Costello, A., 2024. The 2024 report of the Lancet Countdown on health and climate change: facing record-breaking threats from delayed action. The Lancet 404, 1847–1896. https://doi.org/10.1016/S0140-6736(24)01822-1

Santamouris, M., 2020. Recent progress on urban overheating and heat island research. Integrated assessment of the energy, environmental, vulnerability and health impact. Synergies with the global climate change. Energy Build. 207, 109482. https://doi.org/10.1016/j.enbuild.2019.109482

Santamouris, M., Kolokotsa, D., 2015. On the impact of urban overheating and extreme climatic conditions on housing, energy, comfort and environmental quality of vulnerable population in Europe. Energy Build. 98, 125–133. https://doi.org/10.1016/j.enbuild.2014.08.050

Schinasi, L.H., Benmarhnia, T., De Roos, A.J., 2018. Modification of the association between high ambient temperature and health by urban microclimate indicators: A systematic review and meta-analysis. Environ. Res. 161, 168–180. https://doi.org/10.1016/j.envres.2017.11.004

Schwartz, J., Samet, J.M., Patz, J.A., 2004. Hospital admissions for heart disease: The effects of temperature and humidity. Epidemiology 15, 755–761. https://doi.org/10.1097/01.ede.0000134875.15919.0f

Semenza, J., 1999. Excess hospital admissions during the July 1995 heat wave in Chicago. Am. J. Prev. Med. 16, 269–277. https://doi.org/10.1016/S0749-3797(99)00025-2

Shen, Z., Shrestha, S., Chen, C., Mumme, S., Kar, B., 2024. Energy burden aware and thermal resilience informed thermal energy storage system planning for disadvantaged communities. Oak Ridge National Laboratory (ORNL), Oak Ridge, TN (United States), Pacific Grove, California, United States of America.

Steadman, R.G., 1979. The assessment of sultriness. Part I: A temperature-humidity index based on human physiology and clothing science. J. Appl. Meteorol. 18, 861–873. https://doi.org/10.1175/1520-0450(1979)018<0861:TAOSPI>2.0.CO;2

Sun, K., Zhang, W., Zeng, Z., Levinson, R., Wei, M., Hong, T., 2021. Passive cooling designs to improve heat resilience of homes in underserved and vulnerable communities. Energy Build. 252, 111383. https://doi.org/10.1016/j.enbuild.2021.111383

Sunikka-Blank, M., Abdie, D., Bardhan, R., 2021. We need ground space: urban densification and transitional housing in Ethiopia. J. Br. Acad. 9s9, 81–106. https://doi.org/10.5871/jba/009s9.081

Tanabe, S., Nishihara, N., Haneda, M., 2007. Indoor temperature, productivity, and fatigue in office tasks. HVACR Res. 13, 623–633. https://doi.org/10.1080/10789669.2007.10390975

Tham, S., Thompson, R., Landeg, O., Murray, K.A., Waite, T., 2020. Indoor temperature and health: a global systematic review. Public Health 179, 9–17. https://doi.org/10.1016/j.puhe.2019.09.005

The Lancet, 2024. Housing: an overlooked social determinant of health. The Lancet 403, 1723. https://doi.org/10.1016/S0140-6736(24)00914-0

Thompson, R., Hornigold, R., Page, L., Waite, T., 2018. Associations between high ambient temperatures and heat waves with mental health outcomes: a systematic review. Public Health 161, 171–191. https://doi.org/10.1016/j.puhe.2018.06.008

Uejio, C.K., Wilhelmi, O.V., Golden, J.S., Mills, D.M., Gulino, S.P., Samenow, J.P., 2011. Intra-urban societal vulnerability to extreme heat: The role of heat exposure and the built environment, socioeconomics, and neighborhood stability. Health Place 17, 498–507. https://doi.org/10.1016/j.healthplace.2010.12.005

UN-HABITAT, 2010. The Ethiopia Case of Condominium Housing: The Integrated Housing Development Programme., United Nations Human Settlements Programme. Nairobi.

United Nations, 2019. The Sustainable Development Goals Report. United Nations Department of Economic and Social Affairs - Statistics Division.

United Nations, 2023. Sustainable development goals.

Van Daalen, K.R., Jung, L., Dada, S., Othman, R., Barrios-Ruiz, A., Malolos, G.Z., Wu, K.-T., Garza-Salas, A., El-Gamal, S., Ezzine, T., Khorsand, P., Wyns, A., Paniello-Castillo, B., Gepp, S., Chowdhury, M., Santamarta Zamorano, A., Beagley, J., Oliver-Williams, C., Debnath, R., Bardhan, R., De Paula, N., Phelan, A., Lowe, R., 2024. Bridging the gender, climate, and health gap: The road to COP29. Lancet Planet. Health S2542519624002705. https://doi.org/10.1016/S2542-5196(24)00270-5

Visual Crossing, 2023. Total Weather Data - Historical Weather Data & Weather Forecast Data.

Voelkel, J., Hellman, D., Sakuma, R., Shandas, V., 2018. Assessing vulnerability to urban heat: A study of disproportionate heat exposure and access to refuge by socio-demographic status in Portland, Oregon. Int. J. Environ. Res. Public. Health 15, 640. https://doi.org/10.3390/ijerph15040640

Williams, A.A., Spengler, J.D., Catalano, P., Allen, J.G., Cedeno-Laurent, J.G., 2019. Building vulnerability in a changing climate: indoor temperature exposures and health outcomes in older adults living in public housing during an extreme heat event in Cambridge, MA. Int. J. Environ. Res. Public. Health 16, 2373. https://doi.org/10.3390/ijerph16132373

Williams, S., Nitschke, M., Weinstein, P., Pisaniello, D.L., Parton, K.A., Bi, P., 2012. The impact of summer temperatures and heatwaves on mortality and morbidity in Perth, Australia 1994–2008. Environ. Int. 40, 33–38. https://doi.org/10.1016/j.envint.2011.11.011

Workineh, B.A., 2022. The practice and roles of condominium housing for tackling urban problems in the case of Gondar city, Ethiopia. Heliyon 8, e11957. https://doi.org/10.1016/j.heliyon.2022.e11957

Xu, Y., Dadvand, P., Barrera-Gómez, J., Sartini, C., Marí-Dell'Olmo, M., Borrell, C., Medina-Ramón, M., Sunyer, J., Basagaña, X., 2013. Differences on the effect of heat waves on mortality by sociodemographic and urban landscape characteristics. J. Epidemiol. Community Health 67, 519–525. https://doi.org/10.1136/jech-2012-201899

Yenneti, K., Ding, L., Prasad, D., Ulpiani, G., Paolini, R., Haddad, S., Santamouris, M., 2020. Urban overheating and cooling potential in Australia: An evidence-based review. Climate 8, 126. https://doi.org/10.3390/cli8110126

Index